Extreme

Jobs your mother doesn't want you to know about

By Fran Molloy

Australia

Published by Career FAQs Pty Ltd

Published by
Career FAQs Pty Ltd
Suite 76, The Hub
89–97 Jones Street
Ultimo NSW 2007
+61 2 9282 9383
www.careerfaqs.com.au

National Library of Australia
Cataloguing-in-Publication entry:

Molloy, Fran.

Extreme: jobs your mother doesn't want you to know about.

ISBN 9781921106286.

1. Hazardous occupations – Vocational guidance – Australia.

I. Title.

331.7020994

Publisher: Sue Stevens
Author: Fran Molloy
Editors: Darryn King, Alison Edwards and Christine Eslick
Production coordinator: Louisa Veidelis
Production assistants: Jacob Sheen and Bill Birtles
Cover and internal design: Terri Marzullo, H2M Creative Services
Illustrations: Nigel Beard, Ainslie Beard Creative
Desktop publisher: Terri Marzullo, H2M Creative Services
Printed by Paragon Printers Australasia
Advertising sales: Stef Harland

Career FAQs acknowledges the following copyright owners for permission to reproduce their work.

Graduate Careers Council of Australia, www.graduatecareers.com.au

Every effort has been made to contact copyright owners and obtain permission. However, should an infringement have occurred, Career FAQs apologises for the omission and requests that the copyright owner contact them

Disclaimer

The opinions and statements made by people who contributed to this book are not those of Career FAQs. The publishers do not claim to represent the entire extent of the professions included and extreme career choices. The aim has always been to provide a broad overview of the possibilities available.

As such Career FAQs *Extreme* does not purport to be a true and accurate record of the industries included, relying on the voices of those working in the jobs to tell their stories. Inaccuracies may arise as a result of the nature of this book. Users should follow the links to actual websites of organisations to ascertain current practice in these jobs.

 An imprint of Career FAQs

Foreword

A lot of people tell me that what I do is extreme – and I guess skiing at over 200km/h does sound out there. Still, it's not as scary as you may think and it's a whole lot of fun. That's why I do it – for the buzz, for the lifestyle and for the challenge.

I'm sure that's what most people with 'extreme' careers and jobs would tell you. Ask the crocodile farmer, the bomb disposal expert or the war zone journalist why they do what they do. I am 100 per cent sure the answer will be pretty much the same: 'Because I love it'.

Of course, it also helps to be a little crazy. Extreme careers aren't for everyone, but if you like a challenge, and can't see yourself spending a lot of time behind a desk, start thinking outside the square. There are loads of opportunities out there if you know what you like to do and are willing to really look for the perfect job.

You may need to be a little creative. I have never directly made money out of ski racing or speed skiing. In almost 20 years competing at the highest international levels, I've probably made $20 000 in prize money. Most of my income comes from sponsorship and motivational speaking and they both rely on having a good media profile. So I work hard at skiing fast, but I work just as hard on the media side of things.

Here are some tips that may help you find your dream extreme job.

Follow your passion. It sounds simple but most people don't get it. The first step to loving what you do is to work out what you love to do. Think about how you can turn your hobby or passion into a career. I always knew I wanted to ski, especially after I lost my leg to bone cancer at the age of nine.

Set goals and plan to achieve them. My goal when I was ski racing was to win all four races at a Paralympic Games. It took me 10 years but, in 2002 at Salt Lake in the US, I achieved my goal. My next goal was to set a world speed skiing record and ski at over 200km/h, which I did in 2005. Then I wanted to be the fastest Australian skier ever and I achieved that in 2006. Don't lose sight of the end goal – there will always be setbacks along the way.

Develop other skills. Even extreme sportspeople have to sit at a desk sometimes. Remember that you will need other skills to be successful. I spend a lot of time sending emails, attending meetings and doing media interviews. But it's all worth it because I still get to go skiing.

Michael Milton, Australia's fastest skier

Michael is one of Australia's greatest athletes, having won 11 Paralympic medals overall – including six gold. At the 2006 Torino Winter Paralympics (his fifth and last), Michael won a silver medal in the men's downhill.

On a steep track in the French Alps in April 2006, Michael became Australia's fastest skier, clocking a staggering 213.65 km/hr, breaking the previous record (set by Nick Kirshner in 1997) by 1.39 km/hr. He also broke his own world speed skiing record for a skier with a disability for the fourth successive year.

Be creative, tenacious and patient. A standard job is easy to find but what you're looking for may take some imagination, a lot of effort and probably even more patience. Stick it out, it will be worth it in the end!

So, what are you waiting for?

Contents

About Career FAQs

Whether you are just starting out, changing jobs, moving up the ladder or returning to work after a break, Career FAQs books give the inside story on just about any job you can imagine.

Don't end up in someone else's life.
A career choice is really a life choice.
Will Santow, Managing Director, Career FAQs

What makes our books different?

'Career FAQs is Australia's leading careers publisher, with a dynamic new approach to making career choices.'

In each book, employees, employers, recruiters and industry experts tell their stories – so you can really tell if this is the right career for you. You'll find out what is happening right now in your chosen area, get the inside info on the qualifications you need and where to get them, and find out what current employees really love about their jobs.

You also find out how to stand out from the crowd and get that dream job, with our industry-specific résumé, cover letter and interview tips and examples.

Career FAQs is Australia's leading careers publisher, with a dynamic new approach to making career choices.

I started Career FAQs because of my own difficulties in finding empowering, high-quality, up-to-date career information to help me make career decisions.

I wanted to change careers in my 30s, but trying to research a new direction led to a dead end – there was no interesting and reliable information on the qualifications I needed for different jobs, the opportunities available, likelihood of succeeding, or, most importantly, what the work would be like once I got there.

I know that our books can make such choices easier by giving our readers the information they need to find a career that meets their dreams and aspirations.

Our range covers the usual career areas, as well as moving well beyond ordinary categories into careers you may never have thought of. There's a great job out there waiting for you and we can show you what it is and how to get it.

A big thanks goes out to the many people who have taken the time to talk to us and shared their experiences with our readers, as well as to the many eminent Australians who have written forewords to our books, sharing the insight and wisdom that has helped take them to the top.

For a list of current titles, please visit our website, www.careerfaqs.com.au.

With Career FAQs you have all the tools to find your dream job.

Good luck!

Will Santow
Managing Director
Career FAQs

How to use this book

The Expand Your Horizons series of books are designed to suit the theme of the book. *Extreme* has been structured to make it easy for you to find yourself in an adrenaline-fuelled job, faster.

The book is divided into three main parts.

The big picture gives you an overview of the current state of play for extreme careers. It explores the types of job available as well as employment opportunities, and provides an income snapshot of salaries in extreme jobs.

Extreme jobs are profiled in depth in industry-specific chapters – such as race car driver, speed skier, big wave surfer, sky diver, shark handler and crocodile farmer. These chapters include interviews with people in the jobs, employers and information on how you can get started in that line of work.

Ready, set, go for it! You've read the stories, you know how they got there – now it's your turn. Where to start? This section gives you the lowdown on how to go about getting the job you want. Find out how to start your own business, what qualifications you need, how to write a résumé, and more.

To help you navigate through the text and also find information beyond this book, the following icons are used throughout.

 is a snapshot of each interviewee's career pathway – it shows some of the stepping stones along the career path that led to their current position

 directs you to another Career FAQs title that might interest you

 tells you where to find out more about a particular topic or organisation

 provides interesting additional information, which might come in handy!

Many words, phrases, abbreviations and acronyms are defined in buzz words at the back of this book.

 provides a brief definition of a word, term or acronym that appears in bold in the text

 sums up a specific job, including salary, qualifications, number of hours worked, life–work balance and flexibility, as told by our interviewee

 explodes a popular myth about the industry or profession

myweek gives you a day-by-day overview of the typical tasks involved in this job and how they interact with life in general

myday gives you an outline of a day in the job

FASTFACTS provides background information and quirky facts about the job

The big picture

Why extreme?

If you choose a job you love, you won't have to work a day in your life. So what's stopping you from following your dream?

After all, not everyone can stand to work in an office cubicle or behind a cash register, where the closest you'll get to an adrenaline rush is a double espresso for morning tea.

If you prefer some real danger and excitement, perhaps you should think outside the concrete box. Perhaps you should find an extreme career.

These are the jobs your mother doesn't want you to know about.

Our interviewees told us about all the exciting things they do in their work – on land, at sea, even thousands of feet up in the air. It's all in a day's work for these extreme people.

We look at a different job in each chapter, giving you an idea of what the industry's really like. We speak to someone who has made this career their life and get the inside story on how they got there, what they love about it and what the downsides are (if any). We ask them about life–work balance – how hard it is to see friends, family and pets – and what sort of hours they do.

Then we take a look at the industry. For some of these careers, there are only a few people involved in the industry in Australia, and not that many in the world!

Next, we talk to an employer. We ask them who's getting these jobs, how they're doing it, and what you can do to give yourself a leg-up and get noticed.

If you've read this far, chances are you're keen for an extreme career – but you've got a long way to go before you're ready to leap headfirst out of a plane. In Ready, set, go for it, we reveal what qualifications you need to get a foot in the door to your dream job, and how to go about getting them.

You'll learn the best approach and what you can do right now to maximise your chances of being the next extreme worker. Make every day an adventure!

More than 50 million people tuned in to watch the 2006 summer X Games. Summer and winter X Games are held every year in the US.

How extreme?

People in extreme careers can find themselves doing just about anything. There are, however, a couple of ingredients that combine to make a truly extreme career. A generous portion of excitement, the occasional dollop of danger, a smattering of foolhardiness … and maybe a pinch of sheer terror for that extra kick. If this is your dish, you may end up:

- burning rubber on the racetrack

- hurtling down a mountain

- being dumped by the biggest waves in the world

- leaping out of a plane

- swimming with sharks

- fending off the advances of a 10-foot crocodile

- rounding up livestock in a helicopter

- detonating bombs underwater

- tunnelling through the earth's surface

- leading expeditions

- shooting photographs from a helicopter

- doing aid work in a war zone.

What do you need to be extreme?

If you're the sort of person who can be relied upon in a crisis and is not fazed by hard work under trying circumstances, perhaps you're the sort of person who should look for an extreme career. This could be your chance to fulfil a childhood dream like Ian Bobo, former US gold medallist ...

> I don't see myself stopping – ever. I think skydiving keeps us young. I love the fact that we can fly – it's like our dreams as a kid, only better.

Ian Bobo, Skydiving instructor, Pennsylvania, US

What have you got to lose? Besides your dignity, teeth and the feeling in the lower half of your body, you can only gain from living through experiences others would only dream of. It's like what your father always said when you were little – what doesn't kill you makes you stronger. Although this may not be exactly what he meant!

The common denominator shared by all of our interviewees is that they love their jobs with a passion that gets them bounding out of bed everyday, even when there's no milk and the shower's cold. Who wouldn't want to live like that?

Extreme people:

- enjoy living on the edge

- have a love of adventure

- can handle danger

- want to experience things no-one else has experienced!

Ever wondered how extreme athletes do those crazy things? Adrenaline helps. When your adrenaline is pumping you are stronger, faster, and you feel less pain.

What's happening right now?

It's no surprise that more and more people are tired of sitting in stuffy offices and longing for the kinds of thrills and spills you can only find in an extreme career. What might be surprising is that there ARE extreme careers out there, and more are popping up every day!

Some of these careers are so specialised and competitive that you have to be a very talented athlete to have even a chance at making a go of it. Speed skier Michael Milton and big wave surfer Ross Clarke-Jones, interviewed in this book, are good examples of people who have worked their way up the ranks to reach the pinnacle of sporting stardom.

Other careers are experiencing a real shortage of qualified people – such as skydiving instructing and aerial mustering. Now might be your chance to land the job you've always dreamed of.

fyi

Six-time world surfing champion and a pioneer of women's big wave surfing, Layne Beachley has earned around $500 000 in career prize money alone, not including sponsorship.

Extreme women

No-one ever said that only men can be extreme! It might be tempting to think that extreme work is strictly the domain of the alpha male, but women are throwing themselves at extreme careers with just as much gusto and determination as the next guy. And rightly so – brute strength doesn't always win the race.

The extreme women interviewed in this book are:

- Christina Orr – Formula Ford driver

- Isobel Wheeler – skydiving instructor

- Katie Weir – shark handler.

What are the job opportunities?

There are more extreme jobs out there than you can poke a stick at, even if you're an expert stick-poker. In fact, as you'll probably start to realise as you read this book, there are as many extreme careers as there are extreme people.

The jobs we feature fall into four broad sections: *The need for speed, In the wild, Below the surface* and *Getting out there*.

The need for speed

Whether it's down at the F1 or out on the H2O, this section is all about breaking the speed limit. If you like to feel the wind against your face – and an industrial-strength hair dryer doesn't quite cut it – this could be the section for you. We put the pedal to the metal and caught up with a racing car driver, a speed skier, a skydiving instructor, and a big wave surfer.

If this is your thing, maybe you should also consider becoming a:

- bobsledder
- go-kart racer
- storm chaser
- horse jockey
- synchronised pilot.

Warren Faidley is the world's first full-time professional storm chaser. To find out more about him and his work, check out his website.

www.stormchaser.com

In the wild

We're not talking about the furry little things caged up in your local pet store. We're not talking about anything that you have to leave water and carrot sticks out for. We're talking about predators.

You know how foreigners always think we wrestle crocs at shop corners and box with kangaroos just for kicks? Well, there may be some truth in it after all – some people actually do these jobs!

In this section, we take a look at a shark expert, a crocodile farmer and aerial musterer. You could also be a:

- snake charmer
- lion tamer

- bear breeder

- bee keeper

- entomologist

- safari leader

- zoologist.

Below the surface

If you're not afraid of what lies beneath, this is the section for you. Be it underwater, underground or in the depths of a bubbling volcano, these are the kind of jobs that demand a high level of skill as well as the ability to not crack like a fault line under San Francisco. Take a deep breath and check out our interviews with a clearance diver (who works with mines) and a member of an underground blast crew (who works *in* mines).

But the list doesn't end there. Maybe you'd be keen to be a:

- volcanologist

- oil rigger

- lifeboat crewman

- underwater bounty hunter

- pearl diver.

Getting out there

There are some jobs that are just about leaving your area code and your comfort zone.

We spoke to an extreme sports photographer, an international aid worker and an expedition leader, but you could also get out there as a:

- safari leader

- skywriter

- ghost tour operator.

How much can I earn?

We knew you'd mention that sooner or later, so we asked our interviewees how much they earn in their various extreme jobs.

Unfortunately, most of them told us that their career is no way to get rich quick – when was the last time you heard about a crocodile expert becoming an international celebrity worth big bucks? (Aside from *that* guy.)

In any case, even if extreme careers aren't offering the most lucrative pay packages out there, the lifestyles of these wild and wonderful people more than make up for it. It really depends on your priorities, and what's more important to you: an exciting career, or buying your own private island by the time you're 25 ...

As you'll see in our snapshot, our interviewees tell us that their income range can fluctuate from *nothing* to hundreds of thousands of dollars in any given year! The key is to keep in mind that an extreme career is exactly what you make it.

In any case, if you stick at it, maybe you can start charging people for a spin in your V8 supercar or a pat of your favourite croc.

Extreme millionaires include skater Tony Hawk, surfer Kelly Slater and BMXer Dave Mirra.

Age	Position	Approx income ($)
30	Racing car driver	0 – 200K
17	Racing car driver	0 – 50K
31	Speed skier	0 – 40K+
34	Tandem parachute instructor	0 – 2.5K per week
38	Professional surfer	Free backpack – 1 million+
23	Shark dive coordinator	19 per hour
63	Crocodile farmer	25 – 100K
35	Aerial musterer	5 – 50K
18+	Blast operator	40 – 200K
25+	Expedition leader	0 – 50K
28	Extreme sports photographer	10 – 90K
35	Aid worker	1 – 80K

Career FAQs income snapshot 2006

Tradeoffs

What's the catch, you ask? Why isn't absolutely everyone swarming to this action-packed, excitement-filled industry?

Well, apart from there being a lot of wusses out there, there are probably a couple of other reasons.

The very nature of some of these jobs will mean that workers may be away from their loved one for extended periods of time. Add to this the high-pressure environments, the danger and the high-stress situations, and clearly this work is not for just anyone and everyone.

So how extreme are you? Let's find out.

Race car driver

There's nothing like having the wind in your hair and the pedal on the metal – until you get pulled over by the police for doing 100 in a 60 zone.

If you like things at full throttle but don't want to get booked for it, how about a career as a racing car driver?

FAST FACTS

The most expensive crash was at the 2004 Monaco Grand Prix, when race sponsors Steinmetz (a diamond merchant) embedded a $300 000 flawless diamond in the nose-cone of a Jaguar racing car. The driver crashed on the first lap – and the diamond went missing. It was never recovered!

Imagine being at the wheel of one of the fastest vehicles on the planet and getting paid big bucks to burn around a track at top speed. The screech of the tyres, the scream of the engines, the roar of the crowd – beautiful, isn't it?

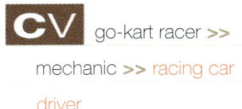
Steven Johnson – V8 supercar driver

Steven is the son of legendary Australian racing car driver Dick Johnson and is a driver for Westpoint Dick Johnson Racing, Australia's oldest car racing team. He believes that there really are only two important things to starting a career in car racing. Firstly, you need the right attitude – the desire to be the best. Secondly, you'll probably need a car. The rest comes with practice!

q&a

Why did you become a racing car driver?

It's in my blood. I've been around cars since I was born and I started racing at six. Because I was brought up around it, I was actually quite good at it so it wasn't hard to decide to make a career out of it. There are a lot of talented people out there who never get a shot – I think I've been one of the lucky ones.

How much does luck play a part in getting a job?

A lot of it comes down to being in the right place at the right time – being around a team at the time that they decide they need someone else, or when a driver leaves.

How young can you start?

Athletes of all kinds can start very early. Like any other sport, you can start racing in go-karts at the age of six. The biggest difference is that racing is a lot more expensive. To play football, you just need a football, a pair of footy boots and a uniform. But karting costs about $10K to $15K a year, just to start out – and you can easily spend double that.

Is it a rich person's sport?

Not always. You do need a lot more money than you would in other sports to get to the top, but if you're at the right place at the right time and catch the eye of the team owner – and you have a lot of talent – then no, it's not just a rich person's sport.

What do you need to do if you want to get into car racing?

The first step is the desire to win, and the desire to be the best possible. You can't get into this sort of thing without having a hunger

to win. You have to keep persevering, keep knocking on people's doors, until the day that you decide to retire.

Do you need incredibly fast reflexes?

You develop these from a young age, especially if you are racing karts, or even playing ball sports like tennis. The more you practise, the more your reflexes improve. There are drivers out there who are not naturally talented but work very hard at it – they do a lot of testing miles and work behind the scenes to ensure they're as good as those who have natural ability.

How much fitness training do you do?

At 6'4" I'm probably the biggest guy in the V8 Supercar field so I have to work very hard to keep my weight and size down as I am racing people who are 20 to 30 kilograms lighter than me. I cycle about 300 kilometres and swim five kilometres per week, and do some weight training in the gym. I run each morning from Monday to Saturday and I do five training sessions in the afternoons as well. All up, I do 11 sessions of training a week.

Is car racing physically demanding?

Yes, you need to force roughly between 40 and 50kg on the gear lever each time you change gear and then roughly about double your body weight, or more, on the brake pedal when you're trying to stop the car. At 160 laps around Bathurst, with roughly 60 gear changes each lap, that's like lifting a 45 kg weight thousands of times in a day with one arm. The heat, the g-forces in the car, operation of the brakes, and the gear is what tires us out. The heart rate of a race driver in a race is usually 130 to 140 beats per minute. And the cars are very noisy and hot – they are certainly not comfortable to drive!

How difficult is it to balance your career with family and friends?

My partner Bree and I had a baby boy, Jet, four months ago. I travel a lot and there are a lot of engagements that we need to do when we are at home as well as race meetings, so it is actually hard to get some time to myself. You get used to having a holiday only between December and February and then you write the rest of the year off. When I'm home I like to stay at home. The last thing I want to do is go away during the holiday time when I have been travelling so much during the year.

What is a typical day like?

No two days are ever the same. Most days I have two fitness training sessions and various driving trials, and then there are all sorts of sponsorship arrangements that have to be fulfilled. We can do very long days – 16 hours and more – getting ready for a race. But the more you enjoy it, the faster the day goes.

Are there any disadvantages, things you have to do that you don't really like doing?

I hate all the politics involved in motor sport and I try not to get involved in it. As a driver I work for myself, contracting to a race team to drive a car. I wouldn't last there if I wasn't doing a good job for them and getting the results both on the track and off. Sometimes you have tough weekends, and long hours. But it's just a matter of persevering through the difficult times.

Are you an adrenaline junkie?

I didn't take up racing because I was looking for an extreme sport – it's just something that I badly wanted to do. I mean, I love my jet ski and dirt bike so I guess the adrenaline is part of it – but I also love my boat and playing golf.

Do you ever worry about the risks involved?

Well, with any extreme sport, there is always a risk that something might happen, and you won't be able to continue in that sport. But even though it is a risky sport, you can't go out there worrying about injury because then you won't put 100 per cent effort into your sport.

What other job would you like to do when you have finished racing?

When I left school, even though I was racing, I did an apprenticeship as a mechanic to give myself a bit more knowledge about motor cars so I had something to fall back on when I stopped racing. Motor sport and racing cars are my life. When I stop driving, I'll move into something else in this industry, whether it be running a team or commentating on racing for a TV network. I know an awful lot about the sport, and that's what I've based my life on, so I certainly wouldn't give that away.

Any career tips for someone interested in motor racing?

If you're interested in racing, you've got to get into a car. Most successful racing car drivers started in go-karts. This is where you learn a lot about racing – setting up the go-kart, race-craft, harnessing opportunities, how to trick people into getting past them and how to race when you've got an ill car. You have to know how to plan to finish in the best possible spot for that race in view of the overall championship. That's what makes champions these days – thinking ahead, not thinking on the spur of the moment.

in brief

V8 Supercar driver

$$$	0–200K
age	30
quals	Automotive Mechanic; International Circuit Racing C (ICC) licence
hrs/wk	50–60
life–work	away from home a lot, working every other weekend

There is a long tradition of car racing in Australia. Australia has the third oldest national Grand Prix in the world, starting in 1928. The first recorded car race in Australia was held in 1901, for three-wheeled vehicles.

Christina Orr – Formula Ford driver

Christina is a 17-year-old racing car driver who turned professional in 2005 and races with the Pure Power Motorsports 'Bad Girls' Team.

She spends a lot of her time training, practising and talking to sponsors and the media.

glossary

Jetsprint race means:

– a competition where a two-person crew race their boat against the clock through water channels on a pre-determined course

Why did you become a racing car driver?

Racing is in my family. My father was a top rally driver in New Zealand and my godfather was a champion at the **jetsprint race** – he bought me my first kart. Dad tried to steer me away from racing. He was hoping I would be interested in ponies.

When did you start?

I was five when I first hopped in a kart, but I was six when I started racing. I finished go-karts at the age of 12. I race Formula Fords now, with a 1600 Cortina motor and in October I am moving over to race Formula Toyotas.

When did you turn professional?

I left school at the end of last year – racing took up so much of my time, I was hardly ever at school. I was offered a spot on the Formula Ford team.

Is it difficult being a woman in such a male-dominated sport?

Most of my friends are boys – I'm surrounded by them. Some guys hate being beaten by a girl, but for most of the guys, on the track, I'm just another racing car driver. I would like to see more girls in racing – the more the better, I reckon. But you've got to be at least equal, if not twice as good as the boys, to be competitive.

As a professional racing car driver, what is your week like?

I go to the gym three or more hours a day. All through the week, I spend a lot of time talking to sponsors and getting the car prepared for the next meeting. I'm waiting for a new deal to go through, for a sponsored drive, so there's a bit of waiting round at the moment. I will be expected to talk to the media a lot.

Are there any things that you don't like about the job?

There is talk at the moment about sending me to an etiquette school, where I will have to learn about talking to the media and get all dressed up. I'm dreading it. I don't really want to go and do that sort of

detour

Find out about more off the
wall jobs in Career FAQs
Weird and Wonderful.

www.careerfaqs.com.au

stuff. I like hanging around racetracks, driving cars. The media side of it really isn't my thing.

What sort of person makes a good racing car driver?

You need to be pleasant off the track – someone who doesn't spit the dummy all the time. You can't carry a big ego off the track. You need to put in an effort to talk to people when they come up and want to chat. You might not be able to talk for very long but you have to be polite. On the track, you can't afford to be too aggressive. You need to show your talent – you're there to win. You need confidence. And on the track, you need to think first.

Do you ever worry about the risks involved?

Once. About 18 months ago, I was involved in a bad accident. A boy I knew was trying to overtake me and came up on the inside of me. The nose of his car hit my back wheel and spun him into the wall, and he was killed. I was spun around but managed to hit the wall backwards. My father had always drilled into me to turn the car backwards and make sure that the gearbox and the engine take the impact in a crash, so I owe my dad a lot. It took a lot to go back on the race track again after that, but it was an accident.

What would you do if you weren't racing cars?

I've always been interested in becoming an embalmer. When I stop racing, I might go on to study that.

Are you an adrenaline junkie? What do you enjoy doing when you're not racing?

I hate flying. I've got a pony at home, but I'm off all around the country racing so I'm not home much. I'm not really into danger. I used to play hockey, but I stopped because it was too dangerous.

I guess racing is a big adrenaline thing. Once when I was racing in the Formula Vees, I was bumped off the start line. I detached myself and kept racing – it was a tight track too, so I was constantly changing gears. I was leading. At the end of the race, I hopped out, walked away from the car and started screaming with pain because the adrenaline had stopped. I had broken my arm and I didn't know – the adrenaline masked the pain.

I do get a lot of adrenaline from what I do, but mostly, I have a lot of fun. Before a race, my parents tell me to go out there and have fun. They don't tell me to go out and win.

in **brief**

Formula Ford driver

$$$	0–50K
age	17
quals	International Circuit Racing C (ICC) licence
hrs/wk	40–50
life–work	away from home a lot, working most weekends

FAST FACTS

In Formula One racing, drivers have automatic gear changing, launch control and traction control systems to assist them. None of this is permitted in V8 Supercars – drivers use their own muscle to shift manual six-speed gearboxes and brake at high speeds.

Race car driver

Careers in car racing

There's a car race of some sort, right across Australia, just about every weekend – and the car racing industry is worth millions.

Make no mistake, though. It can be deadly – they don't call it breakneck speed for nothing. It can be expensive – you're better off playing darts if you aren't prepared to spend some money on karting. And it can be very competitive – there are very few who can make a living from it.

> In Australia, there are only about 50 professional racing car drivers making a living. You basically have to wait for one of them to retire.
>
> Andrew Clarke, Motor racing journalist

The good news is that you can always become one of these 50 – or the 51st – if you get on the right track.

Do I have what it takes?

Although driving talent and fast reflexes are crucial, most in the industry agree that the best drivers have much more than that.

Here are some of the qualities a driver needs to be the best:

- dedication and drive (so to speak) – this means a lot of track training! A professional driver should review the results of car races and training runs to improve driving skills

- self-confidence and belief in your own ability

- knowledge of cars – mechanical skills are a big plus. The driver should be able to report on the car's performance to engineers and mechanics, discuss modifications and test any changes to the car

- ability to look after administrative matters such as licensing, and understand racing rules

- physical strength – professional drivers do intensive physical workouts for two to three hours, six days a week, including running, cycling, aerobics and weights

- strategic approach – know how to finish in the best possible spot

- ability to keep your cool under pressure – too much aggression loses races

- pleasant personality – the best drivers leave their ego on the track

- to be a good ambassador for the sport – you need to know how to handle fans and the media, visit sponsors and give speeches, workshop tours and driving runs

- determination to win.

(It seems that it isn't just about driving really fast after all!)

Who could I work for?

While the motor racing industry in Australia isn't large by world standards, it still employs many thousands of people. The governing body is the Confederation of Australian Motorsports (CAMS), with 52 000 members.

There are around 200 people holding international circuit racing licences, which qualify them to compete overseas, and over 3000 people who hold standard circuit racing licences, qualifying them to compete nationally.

However, there are only around 50 professional racing car drivers employed full time in Australia with the major racing teams.

Most racing teams employ one or two drivers – although some of the bigger teams may have up to six. Many race categories have individual entrants, who fund their own racing.

Racing teams are found on the official websites for each racing category.

John Calamos – Pure Power Motorsport

John Calamos is the owner of Pure Power Motor Sport. He employs drivers – like Christina Orr – on his Formula 4000 race team. His company also runs track days allowing paying members of the public to drive a Formula 4000 racing car – capable of speeds up to 280km/h – provided that they have a driver's licence.

How do you choose someone for one of your racing teams?

Most kids have parents who have been able to fund them through go-karting and they have plenty of experience. It's not a cheap sport, so usually they have been able to afford to put the hours in on the track.

Is having lots of money a prerequisite?

Mostly. There's an old gag in the industry … 'How do you make a small fortune in car racing? Start out with a large fortune.'

With some parents there are huge egos involved and they are prepared to spend enormous amounts of money to give their child a slight advantage to make them win.

However there are still limited opportunities for young people who aren't rich but have a lot of talent to catch the eye of a sponsor.

What personal attributes do good racing car drivers have?

Courage, intelligence and talent. One of the most talented drivers we have is young Christina Orr. She stood out from the crowd from day one.

Any advice for someone who wants to be a racing car driver?

You have to have time behind the wheel to know if you are any good – start in karting and put the hours in on the track. It's very competitive; there are plenty of good drivers out there looking for a break.

Get started!

Still keen to brave the competitive world of car racing? Maybe you should think about taking a driving course – in the USA, there are about 70 schools teaching people to become racing car drivers.

In Australia, the Australian Motor Sport Foundation supports a few select young drivers each year – the rising stars of the racing community – through their Elite Drivers Funding Scheme.

You have to be pretty good to be selected – a committee of racing industry luminaries meets to decide which drivers who've showed promise in local competitions will be offered a place on the program.

Grand Prix is French for 'big prize'. But the biggest single prize in car racing worldwide isn't in Formula One. The National Association for Stock Car Auto Racing (NASCAR), in America, awarded more than US$14 million in prize money in May 2005.

But if you're just starting out, there are other ways to help give your skills a boost. If you have the cash, you can enrol in a program offered by a private racing car school.

But the most common way to become a driver is to work your way up by competing in local clubs – from karting up to the Indy or Formula races. You'll need your own car and support team, unless you can get sponsorship. Unfortunately, most people who are interested enough in car racing to be willing to pay for driver sponsorship often want to have a go themselves!

> All the good drivers who succeed find a way to get funding. They need to show everybody else that they are a talent and have the potential to be a star in the future.
>
> Steven Johnson, V8 Supercar driver

Like most sports, there are many talented young hopefuls starting at a very early age. With kids as young as five speeding around go-kart tracks, it's important to get as much 'seat time' as you can. Join a car club, find a track, and immerse yourself in the racing car culture.

Car racing on the web

Racing schools
www.racingschools.com

Drag race school
www.kenlowe.com.au/DRS1.html

The Australian Motor Sport Foundation
www.amsf.org.au

V8 Supercars Australia
www.v8supercar.com.au

Formula Ford
www.formulaford.com.au

Formula Vee Australia
www.fvee.org.au/drivers

Australian Formula 4000 Championship
www.f4000.com

Australian Karting Association
www.austkarting.com.au

2006 Lexmark Indy 300
www.indy.com.au

NEC Australian Rally Championship
www.rally.com.au

Off-Road Racing in Australia
www.offroadracing.com.au

FAST FACTS

The temperature inside a racing car is often around double the ambient temperature outside – on a 30 degree race day, it can get up to 60 degrees inside the car.

Speed skier

There's nothing quite like standing at the top of a mountain with the world at your feet. It can feel like all you have to do is reach up to touch heaven. There are some people, however, who prefer going down the mountain – as fast as hell.

Speed skiers take their life into their hands when they scour the world for the fastest ski-slopes – and throw themselves down them. The ultimate aim? To ski downhill faster than any other human before them.

FAST FACTS

Skiing isn't a dangerous sport by injury rates – about two people per 1000 are injured while skiing in Australia. Head injuries account for about 10 per cent of all accidents and the industry encourages helmets (though they are not compulsory).

Actually, it probably isn't enough to just say it's 'fast'. Even 'very fast' probably won't do. In speed skiing, the acceleration rate can be similar to that of a Formula One car – from zero to 100 km/h in just over three seconds. As if the idea of wearing a rubber ski-suit wasn't scary enough on its own.

Michael Milton – Speed skier

Comparing speed skiing to regular skiing is like comparing a 100m sprint with a casual stroll in the park. Then there's what Michael Milton does: skiing downhill at more than 200 km/h – on one leg.

Michael, 33, started skiing at the age of three. At nine, he was diagnosed with bone cancer, and had his left leg amputated in a life-saving operation. After a year of chemotherapy, he learned to ski on one leg. In May 2005, Michael became the first person on one ski to exceed 200 km/h, reaching a whopping 210.40 km/h. In April 2006, he beat his own record and became the fastest Australian skier ever setting a new open Australian speed skiing record of an amazing 213.65 km/h.

He's also a ski racer. At the age of 19, Michael came away with Australia's first ever Winter Paralympic gold medal at his second Games. In 2002, at the Salt Lake City Paralympics, he won all four gold medals in his class.

Michael is one of Australia's greatest athletes, having won 11 Paralympic medals overall – including six gold. At the 2006 Torino Winter Paralymics (his fifth and last), Michael won a silver medal in the men's downhill.

What does a speed skier do?

I'm a professional athlete at the moment. I train and compete internationally. I'm starting to diversify – perhaps mountain-lover is a more accurate description.

I climbed Mount Kilimanjaro in Africa last year and soon I'll walk the Kokoda trail.

When did you realise you could make a living out of this?

Although it does make a bit of an income for me now, there were many, many years where it was much more of a cost than an earner. It was 11 years after winning my first Paralympic gold medal that I signed up with the first sponsor who actually put some money in my pocket, though I'd had sponsorship for equipment before that.

Six months after winning four gold medals at the Winter Paralympics, I had to put my racing career on hold and take a job as a ski instructor because I couldn't afford to continue. Fortunately the following year I won a world sport award for my Paralympic achievements, and soon after signed my first cash sponsor – Thredbo Resort.

What did you do to earn money before this?

I've worked in many different jobs. I worked in my family's ski retail business. I've also worked as a coach in ski schools and in resorts in event management (mostly ski events). I've also had jobs away from skiing, such as a leader of a garden crew, driving a ute around and working with a crew on various jobs. I was an education coordinator for the Sydney Paralympic Games.

The only consistent thing about these jobs was that they were all short-term and allowed me to head overseas every summer to train and compete.

Do you still need to work other jobs to survive?

Not at the moment. My income comes from two sources. One of them is sponsorship – I have a great relationship with my sponsors Toyota, and Thredbo, as well as my new sponsor Skins, who make compression garments that improve athletic performance and recovery.

I also work as a public speaker. I talk to people about my experiences on the snow, what it's like to ski at extreme high speed, setting and achieving goals.

Are there any things you don't like about your lifestyle?

Not really. For a long time I didn't have a home. I spent most of my time travelling so I would just keep my stuff in storage at various places around the world. I was always packing and unpacking and I'd sleep in hundreds of differnt beds each year. Now I have my own home, somewhere to come home to, even if it's only for short periods.

What does your job involve on a day-to-day basis?

There are two sides to the job. The first side is the physical side of things, keeping my body in the shape it needs to be in to perform at an elite level. I have between eight and 10 training sessions a week, off-snow. Some sessions are based around a core-strength workout – abdominal and lower back areas.

I do gym and cycling and then I have a lot of recovery-type sessions in the pool. At the moment there is also a lot of hiking, to get ready to tackle the Kokoda Trail.

I try to make as many as possible of the sessions outdoors, because I hate being indoors in a gym – I would much rather be outside going for a bike ride, having fun.

The other side is the administration and organisational side of things. I am very fortunate to have Penni involved, who is my partner and manager. She has taken a lot of that burden off me – she runs my calendar and tells me where I am going and what I am doing.

But there are still plenty of email enquiries to go through, a lot of the information that goes out I have to check through. There is usually three to four hours a day getting through it all, except when I am in training camps or on-snow training when it might be one hour a day.

Is it difficult to maintain relationships with family and friends and your partner when you are away so much?

My sister is also a bit of a ski bum and spends her winters at Thredbo working and her summers in Montana and Alaska in the US, so that's great.

But my partner Penni doesn't travel with me very often. Last year I left in November and didn't see Penni until March.

However, since retiring from Paralympic competition I should be home more. We have a daughter now so spending time at home is very important.

It sounds like you are away for very long stretches at a time.

Yes, skiing is unique because for most sports, you can train at home for most of the year. But with skiing, there is no snow at home when the big competitions are on, so you really are based overseas for entire seasons.

Michael on the slippery slope to success.

What kind of person makes a good athlete?

You need to have a competitive personality, a will to win and to be able to take things further than your competition, sacrifice more, work harder. The most successful athletes are not necessarily the ones who are the most physically talented or the ones that have the best technique.

The people who win are the ones who are the strongest in their mind, and have the desire and motivation to win. In any area of life, if you are prepared to dig deeper, that will put you towards the top of the pile.

How difficult is it to get sponsorship?

Very difficult. I've got four main sponsors at the moment and I have to work very hard to keep them. Because my family owned a ski retail outlet I'm fairly well connected within the Australian ski industry and I haven't really suffered in terms of securing sponsors for equipment.

But even after winning six Paralympic gold medals over a decade, I still didn't have a sponsor until I won the Laureus Award and set a world speed record.

Getting sponsors is not only tough for athletes with disabilities, but also for people in low proflie sports.

What difference does it make having sponsorship?

A huge difference. There are plenty of athletes out there who have won multiple gold medals and don't have any cash sponsors and are working fulltime as well as training fulltime to achieve their goals and fulfil their dreams. I'm really lucky to have several different sponsors who are extremely supportive and who I have a great relationship with, to help me fund my life and fund my dreams and allow me to achieve.

Is it easier to get sponsorship if you don't have a disability?

Yes and no. There is a real misconception in the community about what athletic and sporting success is worth. In many sports, such as swimming, track and field, the football codes, there is certainly substantial financial reward for getting to the elite level.

But 95 per cent of the sportspeople out there are in lower profile sports. They are the ones who, even when they get to the top of their sports, do not necessarily get access to those sorts of financial rewards.

What do you give your sponsors in return?

Sponsorship is certainly a two-way agreement. Both parties must benefit, so I have to really justify anything that I get from a sponsor, whether that's cash or other assistance.

I offer branding on my body during athletic performances. When I set a world ski speed record, that is the time it really pays off for them – that's when it gets on television.

I also offer my services as a motivational speaker, internally for their staff, or for functions that they sponsor, or charities they work with. I offer my name and image in terms of advertising campaigns.

Penni does a lot of the communication with sponsors. We produce a newsletter and regular reports. I do as much as I can to service sponsors because I know how hard it is to get them in the first place.

Check out Michael's website at www.michaelmilton.com

in brief

Speed skier

$$$	Varies considerably
age	33
quals	a little bit crazy
hrs/wk	varies
life–work	challenging

What difference does it make, being an athlete with a disability?

It changes the way I do many things. It is pretty obvious that I've only got one leg. The disability is a part of who I am. It isn't constantly in mind; if I get out of bed at 3am to go to the toilet, I automatically hop there. There is no real need to think about it every day. I think perhaps it is the most important thing to the people I first meet, and the least important thing to my oldest friends who have known me for years and years.

Is it worth it?

Definitely. Being a ski bum is a great lifestyle – I can really recommend it.

my**day**

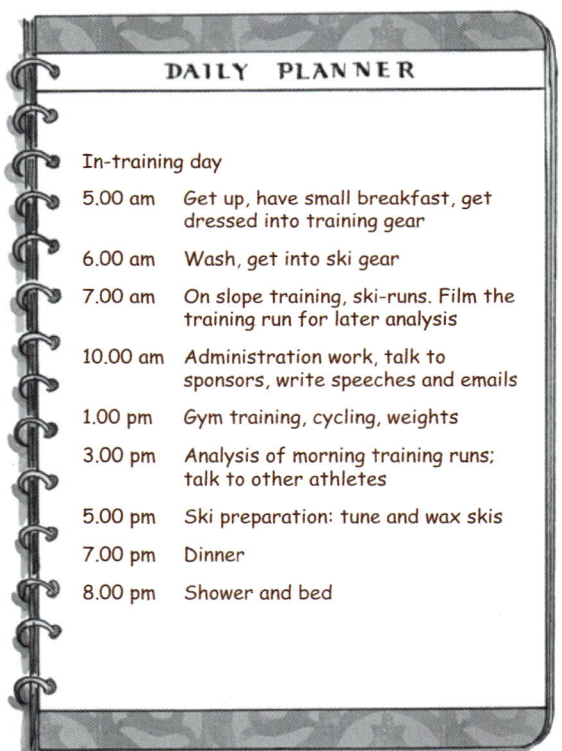

DAILY PLANNER

In-training day

5.00 am	Get up, have small breakfast, get dressed into training gear
6.00 am	Wash, get into ski gear
7.00 am	On slope training, ski-runs. Film the training run for later analysis
10.00 am	Administration work, talk to sponsors, write speeches and emails
1.00 pm	Gym training, cycling, weights
3.00 pm	Analysis of morning training runs; talk to other athletes
5.00 pm	Ski preparation: tune and wax skis
7.00 pm	Dinner
8.00 pm	Shower and bed

Careers in speed skiing

If you love speed and you love to ski, speed skiing is a natural choice for you!

> I ski full time for a living. I feel very fortunate that I have been able to make a living in what I love to do. My speed skiing goal is to find the limit, to ski the perfect run and find out how fast I can go.

Michael Milton, Speed skier, Paralympic gold medal winner

But be warned: it's so competitive that you have to be phenomenally determined and extraordinarily good at it to, well, get anywhere fast, as it were.

Do I have what it takes?

A top-level speed skier is an athlete, and there are several qualities a good athlete needs aside from skills in their chosen sport:

- a competitive nature – the will and discipline to work harder and sacrifice more than anyone else
- a dedicated work ethic
- physical strength – lots of training, on the slopes and in the gym
- strength of mind – in sport, what you're capable of has a lot to do with what you believe you're capable of.

At the elite level, you will of course need a whole other set of skills to do with administration and dealing with the media.

Snow-making was first trialled in Australia at Perisher resort in 1967. Currently, 38.6 hectares of the 1245 hectares of ski fields at Perisher utilise snow-making from 156 guns. Plans for the next 10 years include expanding the area to 110 hectares – almost 10 per cent of the total size of the ski fields.

Who could I work for?

Sadly, browsing the employment classifieds under 'S' won't turn up any jobs for speed skiers. The Fédération Internationale de Ski (FIS) is the main organising body for ski sports. It runs professional speed skiing competitions through the northern hemisphere's winter season.

There are around 200 speed skiers on the international circuit. Very few make a living from sponsorship – most work full-time in other jobs.

Key sponsors include ski equipment manufacturers, ski resorts and other major sporting goods companies, although there are a few other major organisations that may show an interest in sponsoring successful athletes.

Michael Milton, for example, counts Toyota Australia, Thredbo Resort and Skins among his corporate supporters.

Denise Allardice – Senior brand manager

While there are no employers for speed skiers, Denise Allardice is the closest thing – she manages the sponsorship budget for Kosciuszko Thredbo, a large company that runs the Thredbo Alpine resort at Mount Kosciuszko in NSW.

Thredbo is a multi-million dollar village and ski resort – it attracts more than one million visitors each year. Denise plays a major role in deciding which athletes the resort sponsors.

How does Thredbo select a winter sports athlete to sponsor?

Elite athletes seeking sponsorship will usually approach us with a profile, which includes their achievements to date, and their future goals. Then we discuss with them how the sponsorship will benefit both parties. It is important that the athlete recognises that the sponsor needs to obtain benefits, as much as the athlete needs to obtain benefits.

Is it important that the athletes have already made a bit of a name for themselves?

Yes, but that can be a 14-year-old, who has been competing on a junior level for the last three years. It just needs to be somebody who shows promise.

What benefits would you look for from a sponsored athlete?

We ask that they recognise Thredbo as their home resort and that they wear our logo on their clothing when they are competing.

We also ask them to keep us updated with their progress. One of our athletes is currently on the World Cup circuit in Europe. He sends us reasonably recent emails to tell us how he has gone in his latest competition. Then we use that information to inform our customers of what is going on with that person. In a recent email, he told us about having a drink in a bar with one of the World Cup winners – that was also great. For us, it's another talking point, and something that our customers are interested in.

Who are your customers, and how do you keep your customers informed?

We use e-newsletters on a year-round basis for our very loyal customers, who are very hungry for information. Then we have a wider database for direct mail – and we also go out into the marketplace with advertising and so on.

What do you offer your sponsored athletes?

Each of our sponsored athletes has a reserved section on the Thredbo web site. One benefit is that they can list all of their other sponsors in that area – it's great for their other sponsors to be associated with the Thredbo site, so we are also helping to improve the athlete's profile, helping them get other sponsors.

We also assist them financially. Some athletes receive direct payments, for others we might pay their entry into competitions.

We also give sponsored athletes year-round access to our training centre, with the pool, gym and so on. And we give them access to the mountain during the winter, so that's their lift tickets and time on the fields before the public opening times. The platinum pass – year-round access – would normally cost around $1200.

Do you have expectations about the athlete's achievement in their sport?

Yes, certainly. We have had a few athletes in the past that didn't deliver what we expected from them, so their place on the program was given to other athletes.

How long would a sponsorship usually last?

We've sponsored one of our athletes for about eight years now and that is continuing. We usually give an athlete two years to see whether it's going to work, for them and for us.

How much would an individual sponsorship be worth?

From $5000 to $25 000.

Are you able to translate that into any direct business benefits?

Not directly, but it does keep Thredbo in the news. In a small industry like the ski and snowboard industry, it keeps us front-of-mind. And there is the benefit of being a good corporate citizen.

What about lifting Thredbo's international profile?

It really depends. It's a bit of a hit-and-miss program. Currently one of our athletes has been accepted to the Olympic team. If he does very well, that's obviously very good for us. But it's more of a benefit to the industry – it lifts the profile internationally. We would be pretty lucky if the Thredbo logo was beamed into every skier's home in the country.

It is certainly important when someone like Michael Milton – who is climbing Mount Kilimanjaro with Thredbo on his hat and breaking world speed records – mentions that his home resort is Thredbo.

How much pressure is there on athletes to achieve?

I think it's a very personal thing. Some athletes would really beat themselves around the head about it if they lost and feel a real obligation to that sponsorship, while others barely recognise the fact that these bonds exist.

Get started!

To become a speed skier, you need to be very good at going very fast on skis. Sounds simple, doesn't it?

Not many people take their first look at a ski slope and think, 'I'd like to go 200 km/h down that.' Most people are more concerned with staying upright.

Most speed skiers start out as enthusiastic amateurs, then become ski racers, and only later decide to concentrate on Going Really Fast.

Speed skiing is a hard sport to get into. All that specialised equipment is expensive, for a start, but not as expensive as training at the best facilities, which tend to be in places like Les Arcs in France.

While there are no speed skiing tracks in Australia, there are some in New Zealand. Most of the main competitions are held in Europe in March and April.

Want to be a speed skier? Start with slow skiing.

If you think you have what it takes to be a speed skier, you need to be prepared to spend Australian summers at snowfields overseas and winters at snowfields in Australia.

It will be many years before you are likely to earn money in the sport so prepare to work in part-time jobs in hospitality or retail. There is a lot of competition for jobs on the snowfields so you want to make sure you can get work in your supporting field – some qualifications will help, as well as work experience.

One way to set you on the path to becoming a speed skier is to enter downhill competitions on Australian snowfields – and do well in them. Take a look at the Ski & Snowboard Australia (SSA) website to find out as much as you can about qualifying for events.

The steepest ski-able slope in world is a two kilometre icy rock face in the ski resort of Les Arcs, in the French Alps – with a 76 degree gradient. Less than 330 people worldwide have ever travelled faster than 200 km/h on skis – and most of them did it on this precipitous incline.

Success in competitions may interest sponsors and open doors for you to apply for sponsorship. Sponsorship may give you some sort of income, or reduce the costs of equipment and travel, depending on the deal you strike.

Once you've got all that useful experience from downhill skiing – and hopefully some sponsorship and prize money – you can start concentrating on Going Really, Really Fast.

Of course, what you really need to do is get out on the slopes and get as much practice as you can.

Skiing on the web

Ski & Snowboard Australia
www.skiingaustralia.org.au

Olympic Winter Institute in Australia
www.owia.org

Australian Paralympic Committee website
www.paralympic.org.au

Australian Winter Sports Camps
www.awsc.com.au

Snowy staff employment agency
www.snowystaff.com.au

Disabled Wintersport Australia
www.disabledwintersport.com.au

Big wave surfer

Do you like big waves? How about huge waves? How about absolutely humungous waves?

Believe it or not, there are heaps of surfer dudes and babelinis who make a living out of hanging ten on awesome and most excellent waves all around the world. Apparently, many of them even speak normal English.

FAST FACTS

In 2004, Hawaiian surfer Pete Cabrinha cracked the world record for the biggest wave ever ridden, the 70-foot giant Jaws, on the North Shore of Maui, near his home in Haiku.

While there are hundreds of surfers on the professional surfing circuit, most of them compete on smaller waves – which go up to about four metres. However, there are a few professional surfers who surf the big waves – which start at about four storeys. And you thought getting dumped at your local beach was bad.

Surf's up. And up, and up, and up!

amateur surfer >>

professional surfer >>

professional big wave surfer

Ross Clarke-Jones – Professional big wave surfer

Ross, 38, started surfing at age 11 in the coastal NSW town where he grew up. By the time he was 12, Ross was riding waves of up to 20 feet.

While still in high school he started competing in surf contests. Riding big, powerful waves was Ross's strength, but at that time the pro circuit was dominated by slick movers on small waves.

At the age of 19, he finally got to compete in his first professional contest at Waimea Bay in Hawaii. He spent the next 15 years competing on the professional circuit until, in 2001, he had his first big win – he became the first non-Hawaiian to triumph over the 25-foot waves in Hawaii's 'Eddie' competition.

Ross lives for half the year on the south coast of Victoria and the other half in Hawaii.

Do you really get paid to go surfing?

Yes! And I have for 20 years. Quiksilver and Red Bull have supported me for nearly my whole career.

What do you have to do for the money?

These days, it's my own call really. I look up information about swells all over the world and where the surf is likely to be pumping in a week's time. Then I get together a team to try and document it through film and photographs. We fly over the site and work out what we can do there.

We have a purpose-built 22-metre vessel to get out to the waves and then get as much good film as we can. The pictures and footage gets used in surfing movies and company promotions for my sponsors.

Do you only surf big waves these days?

Pretty much, that's my profession.

What's classified as a big wave?

It varies from person to person. But I would say that a big wave is 15 to 20 foot. You start calling it a big wave if it's over 15 foot [around 4.5 to six metres]. That means it's got a 30-foot [around nine metres] **face**. Then it can go into 40 foot [around 12 metres], which is an 80-foot [around 24 metres] face – huge waves. That's another realm.

Has big wave surfing changed with the introduction of jet skis?

Definitely. I have ridden waves up to about 30 feet, maximum, at Waimea Bay (Hawaii). Those waves are the fastest a human can paddle onto with a surfboard. After that, the waves just move too quickly. You need to be propelled by something, either a boat or a jet ski. We've been doing that for the last 10 years. Using jet skis completely changes the way you look at the ocean. There are waves you can't paddle out to which are kilometres out to sea. Jet skis change the whole playground completely.

What's the biggest wave you've ever ridden?

I rode the 'Biggest Wednesday' at **Jaws** in 1998 (on the north shore of Hawaii's Oahu). That day had waves of 45 feet [around 14 metres] – which works out to be about 90-foot [around 27 metres] faces. Me and Tony, my surfing partner, drove right in front of a wave and both of us were smashed – me on the surfboard and him on the jet ski. The wave just swallowed us.

What happens when you get hit by a wave that size?

You used to get pummelled for a couple of minutes in a really bad wipe-out. But now, lifejackets cut that down to about 30 seconds. You are completely thrashed for that 30 seconds, but at least you don't have to try and swim to the surface – you just float up. Some lifejackets have a one-minute oxygen tank. I don't use them – after 20 years, introducing something like that would change my whole thinking, worrying about when to use oxygen. When I'm underwater getting thrashed, I actually relax.

How do you find the life–work balance in this job? Is it hard to catch up with friends and family and maintain a relationship?

I'm in Victoria at the moment, but I could leave at any time, so I'm always on the run. I'm based in Victoria from March until October, and then in Hawaii, on Oahu the rest of the year. That has been my standard schedule for 21 years. This will be my 21st winter in Hawaii. I spend about four months there. I have two children who live with their mother – every winter they come to Hawaii and they spend school holidays with me, so that works really well.

glossary

Face means:

– the forward-facing portion of the wave where riding usually occurs.

Jaws means:

– a beach on the Hawaiian island of Maui, famous for having the biggest waves in the world.

find out **more**

Check out the pics of Ross Clarke-Jones in action at www.redbull-photofiles.com

Ross says his job is one of the best in the world.

What's it like for young surfers coming into the professional business today when there are a lot more sponsorship opportunities around?

A lot of the most talented young surfers these days get scooped up at an early age and are paid enormous sums of money before they actually do anything. I think it may be damaging if it doesn't work out, because they have managers pampering them and they don't learn anything for themselves. It was a real struggle to survive when I was going through so you appreciated every moment.

Years ago, it was almost impossible to earn money outside of the professional competition circuit. But now there are many forms of sponsorship – there are pro aerial surfers, pro soul surfers, pro adventure surfers, and pro big wave riders.

Would you be able to do this without sponsorship?

I'd like to think that I would, but the reality is that without the sponsorship I'd need to have a second job, which would strip me of the freedom to drop everything and chase swells.

What do you like about the career you're in now?

My sponsors support me while I'm off travelling the world basically all year filming, visiting surf shops and hanging out with different groups of surfers around the world, promoting Quiksilver and so on. I have the freedom to surf wherever I like. The money is as good as a professional tour, but I don't have to travel to all the different contests. I create my own tour.

It's one of the best jobs in the world. I can't think of a better one.

in brief

Professional surfer

$$$	from a free backpack to over a million dollars a year
age	38
quals	natural talent and a high level of commitment
hrs/wk	all the daylight hours that the surf is good
life–work	constant travel and always on call

Careers in big wave surfing

If you are one of the few people who can look at a wave with a 20-metre face and think 'Cool!' rather than 'Aaaaargh!' then maybe big wave surfing could be your thing.

Making your way in this sport takes talent and experience as a surfer, plenty of access to those remote and often difficult-to-reach places where the big waves really pump – and a whole lot of bravado.

The seven-storey waves that smash across places like Hawaii's Oahu North Shore are not an easy ride – you need be very strong and fit to handle them.

> When it comes to conquering big waves, it's an individual matter of desire ... I think it comes down to mental ability and the ability to relax under extreme duress. You've got to love the adrenaline and the challenge to face and overcome your fears. You've got to be comfortable in really uncomfortable situations.
> Layne Beachley, World champion surfer

But whether you're a big wave surfer or a regular surfer, there are two main ways that you can earn a living just from surfing.

The first is to break into the professional surfing competition circuit and compete for the prize money. This can be a precarious existence and you need to win small competitions frequently or big competitions every couple of years to make a living from it.

The second way is to get sponsorship from a company prepared to fund your travel and competition entry in return for association with their brand, perhaps some advertising appearances and attendance at promotional events.

If you are among the world's best, a sponsor may not expect constant appearances on the tour but could allow you freedom to travel outside the main routes.

Do I have what it takes?

This is a career like no other. You need technical proficiency. You need to be extraordinarily fit. You need to be physically powerful. You need to be very passionate, committed and to believe in yourself.

> For me, extreme sport is not just about competing. It's about the complete satisfaction of conquering whatever discipline you choose.
> Ross Clarke-Jones, Professional big wave surfer

A good surfer needs to be:

- fearless – because you will get dumped
- tenacious – because you will continue to get dumped!
- at peak athletic fitness.

Apart from this, if you really want to get anywhere professionally, you must also be able to convince sponsors to believe in you and want you to represent their brand.

Who could I work for?

Pro surfers usually aren't 'employed' as such. Instead, they attract the attention of a sponsor, usually a surfing-related company which signs them up on a contract to represent them either by joining their surf team or promoting their products.

Really, the most important part of going from an amateur level to becoming a professional is the ability to win surfing competitions. We spoke to one of Australia's most senior judges for some tips.

Glen Elliot – National judging manager, Surfing Australia

Glen Elliot is the current national judging manager for Surfing Australia and a former competitive surfer. After 25 years of service he is now a life member of the Palm Beach Boardriders Club. He tells us about the level of commitment and talent needed to become a professional big wave surfer.

q&a

What path would you take through the amateur ranks to have a chance at becoming a pro surfer?

Australia has a unique system that creates a certain pathway to pro surfing. I'm not saying it's the only way but it has contributed to our dominance of world surfing in the past. We have local board-riders clubs where kids can start at an early age and learn to surf at a competitive level.

There is no age limit – some clubs have great-grandfathers surfing with their great-grandkids. Three generations is common. Once you're a member of a local club, you're usually there for life.

It's at club level that competitive tenacity really takes hold. You compete in your age group – from the 'Micro-Grommets' (about seven or eight), then 'Grommets' (up to about 14), and then you have your juniors. Once you do well in your age group events, you have the chance to compete at regional events and state titles.

How would you know if you were likely to make it?

Most pro surfers are usually well on the way by the time they are 16. They are offered sponsorship pretty early on. A couple of Australia's top 16- and 17-year-old surfers are on contracts and make between $60 000 and $70 000 per year.

What kind of person makes it as a pro?

A top professional surfer is like any elite athlete. They have to be very competitive and have a massive amount of tenacity – they will never give in. They also need to have a lot of courage. A lot of the contests held throughout the world are in really challenging waves – the best pros go in and attack the waves and don't let fear beat them.

People skills and ability to handle the media is something that they usually learn as they go along, and Surfing Australia has a number of really committed people in the High Performance Centre who give them a hand with that sort of thing.

What advice would you give to someone really keen to become a professional big wave surfer?

Join your local surfing club and start entering competitions. You'll soon find out if you have a chance to go anywhere with it.

Get started!

Turning pro is not easy. Most professional surfers start by winning their way through the ranks in amateur surf contests and getting recognition that way. Local boardriders associations can be a great source of advice and mentoring.

You need to work your way through the surfing hierarchy, so it helps if you start young and live near a beach with decent surf.

Talent, luck, passion and hard work all play a part.

Of all the surfers who try out, only a few make it to the top. The top five on the circuit can make millions from sponsorship deals, but the prize money isn't all that huge. The next level down on the WQS (World Qualifying Series), most pros make barely enough to pay for their travel. A lot of sponsorship tends to be for product – a free board or something. There would only be 50 or so in Australia actually making a living from surfing.

Jak Carroll, Course coordinator, Diploma of Sport Management (Surfing Studies), Southern Cross University

≡FAST FACTS

In 2001, Ross Clarke-Jones was the first non-Hawaiian to win the Quiksilver in Memory of Eddie Aikau Big Wave Invitational event. His reward for more than 14 years learning to master the huge waves of Hawaii's Waimea Bay was a cheque for $50 000.

Courses and study options

There are not many courses that can prepare people for a surfing career, although there are many private surf schools around.

Some TAFE colleges offer surfing-related courses, such as a Certificate IV in Sport and Recreation – Surfing (Melbourne TAFE).

Two Australian universities currently offer surf-related courses – Southern Cross University has a Diploma of Sport Management (Surfing Studies), while Edith Cowan offers a three-year Bachelor of Science (Surf Science and Technology).

You should live, eat, breathe and sleep surfing. But you need something else – the drive and mental discipline that only elite athletes achieve.

FAST FACTS

The surfing community is divided over the use of jet skis and helicopters to access big waves that are moving too fast for surfers to paddle onto. Since the 1990s, surfers have been towed into a breaking wave by a partner driving a jet ski or helicopter with attached tow-line. Critics argue that surfers can get into situations they won't survive.

You need to be able to relax *when* – not *if* – you are held under water for over a minute by the crushing white water that comes after a huge wave. You need to be very, very fit and strong. Experienced surfers call this 'most gnarly'. Normal people call this freaking terrifying.

> I don't think age has much to do with it. It's more a mental state than anything and how much you can keep your body in tune. Professional surfing takes a lot out of you, more so the travel than anything else, and it's just a matter of learning to make yourself comfortable, but keep your Ferrari – your body – fine-tuned, well-lubed and prepared to deal with the pressures of competition and travel.
>
> Layne Beachley, World champion surfer

Working part time in a surfing-related field – such as in a surf shop – can be a good way of earning money while keeping abreast of the industry and opportunities for sponsorship.

If you think you have what it takes to be a pro surfer, hit the water right away!

Surfing on the web

Surfing Australia
www.surfingaustralia.com

Association of Surfing Professionals – World Tour site
www.aspworldtour.com

Southern Cross University – Diploma of Surfing studies
www.scu.edu.au/schools/essm

Australian Surf Industry Training School
www.asits.com.au

International Surfing Association (ISA)
www.isasurf.org

Board Sport Jobs – Jobs in the US Surf industry:
www.boardsportjobs.com

Skydiver

For some people, it's scary enough getting into a plane – let alone diving out of one at 5000 feet, with little more than a sheet and a piece of string between you and a nasty splat.

But some people get off on jumping headfirst from a plane – and coaxing other people off the ledge too!

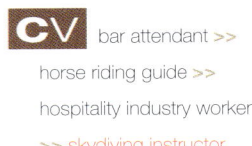

bar attendant >>

horse riding guide >>

hospitality industry worker

>> skydiving instructor

Falling skydivers often reach a top speed of 200kph, though a higher jump can reach a greater speed. This is due to terminal velocity – the top speed possible for a falling object. Depending on a number of variables including wind resistance, air density and surface area, a falling object will stop accelerating at some stage because it reaches a speed where the force of gravity is equal to the opposing force of air resistance.

Isobel Wheeler – Tandem parachute instructor

Isobel is 34 and started skydiving as a hobby when she was about 20. She became passionate about the sport and continues to compete in international freefall flying. She holds two world skydiving records and had completed more than 3000 jumps before starting work as a tandem skydiving instructor six years ago. She fell in with Paul's Parachutes in Cairns about a year ago.

q&a

What do you do in your job?

I'm a tandem instructor, freefall photographer and freefly coach. I take people for tandem skydives, I take photos and videos of people's jumps and I also teach a style of flying known as freefly.

What are the usual employment arrangements for skydiving?

For tandem instructing and camera work, I'm a contractor with the parachuting company, which is a private company. For teaching freeflying, I am usually an individual contractor either to a skydiving body, a private company or an individual.

What do you like best about your occupation?

It's a pretty good office – it doesn't have four walls. The energy in this job is really good. Everyone is always on a high. And you have a lot of freedom. You're not hanging around a computer desk all the time.

Are there any things you don't like about your lifestyle?

I work very odd hours. We start usually at about 7.00 am and sometimes when you are really busy you might still be working at 7.00 pm. It can take a toll on your body after a while. A lot of people get bad backs or shoulders from this sort of work. I try to keep flexible by doing yoga and lots of stretching.

Do you ever get scared?

I don't think so. The first jump I did was in 1992 and I thought, wow! What an amazing thing to do. I've done more than 3500 jumps since then and I still absolutely love it.

Have you ever been in a life-threatening situation?

I've been pretty fortunate – so far I haven't had anything really scary

happen to me. I have seen training videos and so on which show you what can go wrong. I think if you take the time to talk to your passenger you can eliminate panic as an issue. There is always an element of risk, but I am very safety conscious and I think that is partly the reason that I haven't had problems, touch wood.

What kind of person makes a good skydiving instructor?

You have to be able to remain calm, no matter what. Never make rash decisions – you need to make safety your priority. You also need to have good communication skills, and to be able to read people in quite a short amount of time and work out what they need.

How did you become an instructor?

I waited until I was about 30. (Most instructors are over 25 – I think this is because you need to have a little bit of life experience.) I sort of fell into it because I was so involved in skydiving as a sport. I used to earn a bit of money as a part-time **camera flyer**. A couple of years after that I decided to move to instructing on a full-time basis.

How difficult is it to get work?

At the moment there's a shortage of instructors, so if you're qualified, it isn't all that hard to find a job. Most people don't start out deciding to instruct, they get involved in it as a hobby. I'd done about 3000 jumps and had been in a lot of competitions – I'd already won two World Championships in freefall flying when I became an instructor.

It is quite a long road to qualify and you have to be keen. No-one does it for the money.

Is it difficult to balance work and family life?

A lot of people who are instructors have families and it can work really well – you can be flexible with your hours because you are a contractor. My partner is a canefarmer and works crazy hours so that can be a bit tricky, especially when he lives about one-and-a-half hours away.

What do you wish people had told you before you started working in this job?

I was pretty well informed because I had been around the industry for such a long time. But something you do need is to be a good money manager – it can be quite seasonal work, which means you don't have a regular income.

What training and qualifications are important?

The industry is quite tightly regulated so you can't be licensed as an instructor until you have quite a lot of experience. There are all sorts of jump requirements for every step. For example, up to 50 jumps you can only freefall on your belly, after that you can start doing a bit more. You have to have 200 jumps to be a camera flyer.

You can't be an instructor with anything less than 600 jumps, a packing certificate and approval from a chief instructor.

glossary

Camera flyer means:

– someone who takes pictures of other people's skydives.

find out more

www.paulsparachuting. com.au

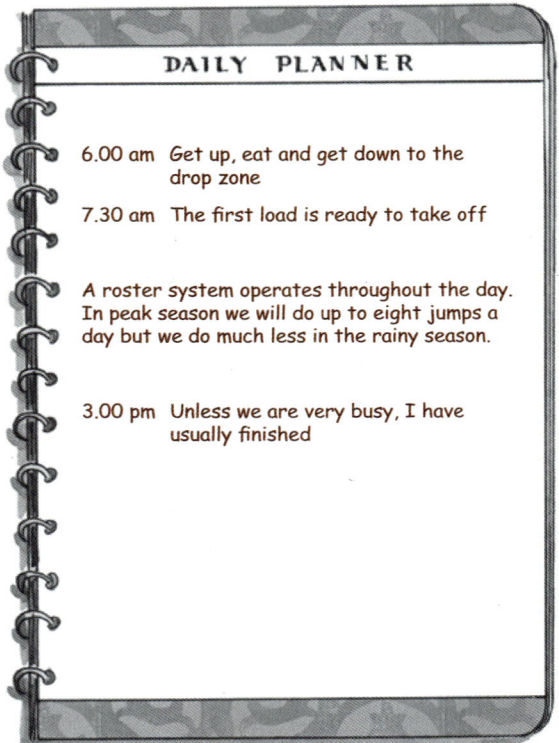

my**day**

DAILY PLANNER

6.00 am   Get up, eat and get down to the
drop zone

7.30 am   The first load is ready to take off

A roster system operates throughout the day.
In peak season we will do up to eight jumps a
day but we do much less in the rainy season.

3.00 pm   Unless we are very busy, I have
usually finished

ⁱⁿ **brief**

Tandem parachute instructor

$$$	0–2.5K per week
age	34
quals	Minimum 600 jumps, plus numerous other specialised qualifications
hrs/wk	3–80
life–work	very flexible

Careers in skydiving

If you've never done it before, it's possible that the thought of jumping out of an aeroplane gives you goosebumps. You either love the idea of it or you hate it. Many skydivers admit that they started doing it in an effort to overcome a fear of flying and heights.

Skydiving is the ultimate extreme sport. Unlike many other extreme sports, however, you don't need to go through years of training to experience the full rush.

An opportunity to take a tandem jump with a fully qualified instructor is readily available. Given that it includes a plane trip, a tandem skydive is relatively inexpensive.

For many people, the thrill of their first skydive is enough for them to want to make it their life's passion. While you don't go into skydiving to

make lots of money, it's certainly a career that's not short on excitement.

A skydiving instructor teaches people to skydive. Before making a skydive, a parachuting student must be taught by a certified instructor. People who skydive can choose either to take a skydiving course and get a licence allowing them to jump unassisted, or they can give it a try with a tandem jump.

FAST FACTS

Conversation in freefall is virtually impossible. The wind is too loud.

Before the first jump, the instructor explains the normal and emergency procedures for all aspects of the jump and demonstrates how the equipment works. Often, instructors use a training video for this part of the course.

There are three types of skydiving training: static line, freefall, or tandem. Later, we will look at these in detail.

In Australia, the majority of skydiving instructors are involved in tandem jumps – allowing beginners to skydive with an experienced instructor.

The student and instructor both wear a harness, but only the instructor wears the parachute. The student's harness is attached to the front of the instructor's harness and the two of them freefall together for 30 seconds or more, open together, and land together under the one parachute.

Do I have what it takes?

In this sort of job, people put their lives in your hands – and your parachute – every day. Not everyone is up for that sort of responsibility, but there are a few qualities that make a good skydiving instructor:

- an authoritative but reassuring presence – you need to convince people they can jump out of a plane and live to tell the tale!
- the know-how and experience – you need to make sure they actually do live to tell the tale …
- a cool head in an emergency.

And, whatever you do, don't forget your parachute.

Who could I work for?

There are plenty of drop zones in Australia you can drop into if you're serious about becoming a skydiving instructor.

Glen Bolton – Chief skydiving instructor

Glen Bolton is the chief instructor at Paul's Parachutes in Cairns. We talked to him about the sort of person that's really good at making people jump out of planes.

When you employ an instructor, what skills or attributes are you looking for?

My priority is that the people I employ are safety-minded. They need to present a really professional attitude and demeanour. They need to be someone that you would be happy to trust your life to, because that is precisely what our customers do every day.

The instructors have to be team players and need to support each other. The hardest part is getting your passenger out the door of that plane. You will have a whole collective of skydivers, talking and assimilating with the person on the way up and they need to give out this feeling of confidence together.

What don't you want in a tandem instructor?

I don't want someone who is gung-ho and full of themselves. That sort of person is just going to be dangerous. That's why it's very rare for me to employ someone that I don't already know through the skydiving world.

They must have a lot of experience, not just in the number of jumps but the range as well. If I didn't already know them – and, like I said, that's rare – I would look at their credentials.

You've got to have leadership qualities. They must be good at what they do. Isobel Wheeler is a world champion and a darn good skydiver, so she is very confident and that confidence is infectious.

People pay for the thrill of skydiving, but they need to be assured that they will be safe. Often that calm confidence is something that you won't find in a younger person – I have never employed an instructor under 26, for example.

What training do they have to do?

It's a fairly long road to qualification and we have something of a crisis of retention of instructors at the moment. There is certainly a shortage of tandem instructors in Australia.

The training involves going through a series of licence requirements, including a packing certificate qualification. An instructor needs a minimum of 600 jumps. It's quite a long haul.

For example, the final stage of the instructor's course is two days in the classroom, plus a number of written tests. They have to do an oral exam with a panel of three examiners, as well as 10 tandem skydives as an instructor on which they are tested, and other requirements like forward leaps and backward leaps.

What tips would you give someone wanting to become an instructor?

Start jumping. But you need to have the right mentality and the right heart to carry it off, as not everyone will suit this career.

Get started!

The Australian Parachute Federation is the official certifying body for parachute licensing in Australia.

There is an extensive list of parachuting licenses and complete details of requirements for each licence is updated on the APF website.

There are three types of training available – static line, freefall, or tandem. In Australia, the majority of first skydiving jumps are done as tandem jumps, but this is not part of a course that will help you obtain a skydiving licence.

Static line is the traditional military parachute training method. Students spend a number of hours in on-ground training before being taken up for their first jump. They fall away from a plane above 3500 feet, and the main parachute is opened after two to three seconds by a 'static line' attached to the aircraft. After several more static line jumps students can do their first freefall jump.

There are two freefall training methods – accelerated freefall and tandem-assisted freefall. The accelerated freefall method involves extensive ground training and then students do a 40–50 second freefall from 12 000 feet with two instructors, who hang onto the student from when they leave the aircraft together until the student pulls their own ripcord at about 4000 feet.

In tandem-assisted freefall, the first three stages of freefall training are completed as tandem jumps. This method takes longer than the previous two so can be more expensive.

For all training methods, you progress through a number of jumps with progressively longer delays before pulling the ripcord and then

meet other requirements, such as being able to pack a parachute, before qualifying for an 'A' licence.

You can work through a long list of licences to obtain qualifications that let you start work as a professional rigger, camera operator and eventually skydiving instructor.

To obtain a Parachute licence A, a skydiver must have completed at least 10 stable free falls, made 10 landings within 25 metres of the target centre, completed packing a main parachute and passed a written exam.

FAST FACTS

The majority of skydiving fatalities these days are a case of operator error. In 2002, about 40 per cent of all US skydiving deaths were due to the skydiver making a mistake in landing, 21 per cent were due to them colliding with another skydiver in freefall or under their parachute and 12 per cent of deaths were due to the skydiver not responding correctly to a 'survivable equipment malfunction'.

How much will it cost?

If you think you would like to be a parachuting instructor, you need to start saving up your money to put yourself through all the jumps and courses you will need to qualify.

It will be some time before you're likely to earn money in the sport – so prepare to work elsewhere in the meantime.

Parachute schools estimate that you will need to do about 200 jumps before you have enough experience and qualifications to 'neutralise your jump bill' – for example, getting free jumps in return for taking films or photographs of new jumpers.

First jump prices in Australia range from $150 to $400 per jump. However, jump prices are much lower when you are completing a training course.

The overall cost usually depends on the training method you choose, With freefall and static line training methods, for example, the cost of successive jumps decrease as less supervision is required.

Once a skydiver has qualified for their 'A' licence and they own and pack their own gear, jumps can drop to between $30 and $50. Still, it is very expensive to obtain the necessary number of jumps to get qualified. But most parachute instructors will tell you it's worth every cent!

Then there are a series of other licences and exams to qualify as an instructor.

To become a parachute instructor, you need to have very good parachuting skills yourself. These are developed over time. Many instructors also have plenty of experience in skydiving competitions, both national and international.

Competitions are held regularly and events include categories like freefall style skydiving, accuracy landing, formation skydiving and canopy formation. Newer and more radical events include freestyle and skysurfing.

Many event categories allow skydivers with less than 200 jumps to compete in a team provided that they have attained certain qualifications.

Check the APF website (see *Skydiving on the web* at the end of this section) to find out about what events are held in your local area and how you can qualify.

If you haven't already done a parachute jump, first you'd better try it and see if you like it!

No-one will claim that skydiving is a safe occupation. But accidents are much less common than is widely believed. The Dropzone unofficial list of skydiving fatalities lists 69 fatalities worldwide in 2004, and 59 fatalities during 2005.

In Australia, the minimum age to take part in a tandem skydive is 14 years. Anyone under 18 years must have the written consent of their parent or guardian. You must be 18 to do a solo skydive.

The maximum weight for the passenger in a tandem skydive is usually 95 kilograms.

Skydiving on the web

Australian Parachute Federation
www.apf.asn.au

International Parachuting Commission
www.fai.org/parachuting

Australian Skydiving Magazine
www.skydiver.com.au

Skydiving Magazine
www.skydivingmagazine.com

Sydney Skydivers
www.sydneyskydivers.com.au

Skydiving Australia
www.skydivingaustralia.com.au

The Parachute School
www.skydivingassociation.com.au

Shark handler

Just when you thought it was safe to go back into the water. Fancy your chances in a pool full of sharks?

Feared by swimmers, sailors, fishermen, moviegoers and other fish, sharks are the quintessential predators of the deep.

Our fascination with them is so strong that films and folklore about sharks will always have a market; people flock to aquariums just to see them, and tourists will put up thousands to swim with them.

So what's the deal with sharks?

 FASTFACTS

A report from the University of Florida claims that shark attacks dropped in 2005 partly because people are fighting back more, often by hitting sharks. Attacks have been on the decline for five years worldwide, since reaching a record high of 78, 11 of them fatal, in 2000. Search the Internet for the *International Shark Attack File* for more detail.

Extreme

Katie Weir – Shark dive coordinator

Katie Weir is 23 and works part time as a dive leader at Oceanworld, in the beachside Sydney suburb of Manly, while studying for a law degree. It's her job to join paying customers in the giant aquarium with grey nurse sharks, stingrays, sea turtles, wobbegong sharks, moray eels and other friendly marine creatures.

q&a

What does your job involve?

It's mostly teaching. My job is to take people through the shark tanks. About 90 per cent of the people we take into the tank have never been scuba diving before. We start from the beginning, take them through all the skills and teach them to breathe underwater using the regulators and scuba gear before we go into the tank with the animals.

In a typical day how many times would you go into the tank with someone?

Usually just two dives a day, but on busy weekends it could be up to four times. We're only in the water with the animals for about half an hour, but the whole process of meeting the people, taking them on a tour, teaching them the skills they'll need and then having a pool session takes about two-and-a-half hours all up.

How many people would you take at a time?

If they are all certified divers, we can take five at a time, but if we're teaching people how to dive, there are two instructors for every four people.

How did you get into this job?

Through contacts – I knew the manager through scuba diving. I volunteered to work here for a few months before they gave me a paying job here. They want to be sure that you can handle the animals and take clients diving in the tank and so on. Most people working here started by volunteering on the weekends and hoping that a position would open up.

When did you first start scuba diving?

I've been diving since I was 12, when I did my open water diving certificate. Then I went through and did additional dive course levels

above that, advanced diving, then rescue diver and a few other courses. I really love being in the water. My family live near the beach, my dad's always been a diver and we have dive equipment all around the house. I tagged along when my older brother did a dive course. I was the youngest on the course.

Have you dived with sharks in open water, rather than the aquarium?

Yes, I've done a couple of shark dives while I was in Fiji where you actually feed the sharks but it's also quite common to see them when you are just diving around Sydney. You see grey nurse sharks, Port Jackson sharks, wobbegongs and so on everywhere when you know where to look.

What did you think the first time you were underwater with a shark?

The first time I saw a scary-looking shark was on the reef in Fiji. We were down about 15 metres diving on the reef on my first day diving in Fiji – this was only about three months after I learnt how to dive – and suddenly I looked over my shoulder and there was a two-metre black-tip reef shark just following behind us. It was an awesome moment. I was excited and maybe a little bit anxious – but then my brain kicked in and I realised it was all right, it was just checking me out.

Do you feel comfortable diving with sharks now?

Definitely. I never had a phobia of sharks – I swam at the beach all the time and don't really worry about it. Now, after diving here so much, I don't worry about them at all. When I'm in the tanks so much with them, they're just so graceful that you don't really even think about them as sharks.

Have there ever been any worrying moments?

About a week ago, our big wobbegong shark, Brutus (who is a bit of a cranky shark), was coming in towards one of the clients that I had taken into the tank. I popped up next to the client to warn him that Brutus was coming along in front of him, and just as I did, I actually knocked him on the tail and he spun around and gave me a bit of a scare. He didn't snap at me but he wasn't happy. I'm usually more worried that something is going to happen the clients than to me.

Is there a certain way you should behave around sharks?

You should have respect for them – don't move towards them, and keep your hands away from them. I have had clients try and reach out and touch the sharks, try to get them to brush against their head as they cruise on over the top ... which is not a good idea. The sharks that we have are very used to having divers in the water so they don't mind too much, but there are definitely ways you should act around sharks. We've never had any incidents.

What do you like best about this job?

I love doing this job. Most of the other staff members are full-time, though there are a couple of part-timers. I love the chance to go diving all the time. You also meet some incredible people. The other day, I took a girl in the tank who was mostly blind, she could only see about two inches in front of her face. Meeting inspirational people like that who get out and do these kind of things is one of the best parts of the job. Being with the animals is also very cool.

What about the downside? Are there any things you don't like?

On a cold winter's day the water does start to get a little bit cold – especially when you have an early start on a weekend. Towards the end of the third dive, after half an hour in the water, it can get uncomfortable – you really need a good wetsuit. Teaching some people how to dive can be hard work. And sometimes, we have larger people that we really have to stuff into the wetsuits. Not the best part of the job!

What advice would you give to someone who was really interested in becoming a shark handler?

Diving skills are important. Do as much diving as you can and get lots of experience. Volunteer work is also a great way to get into it. These jobs are quite hard to get into unless you know someone but being involved in the diving industry will help. Just being involved, helping out in the tank and so on is a really good way to start out.

What sort of person would make a good shark handler?

Someone who is outgoing. While there is a lot of diving involved, your people skills are more important because you're teaching people as well. It can be quite a challenge taking someone with no scuba diving experience into a tank full of sharks.

Should you have an instructor's certificate?

You need to be a divemaster or an instructor. I'm a divemaster but most of the people here are instructors.

What's the pay like?

As a casual, I'm paid $19 an hour, which is great money for doing something I love.

in brief

Shark dive coordinator

$$$	$19 per hour as a casual
age	23
quals	scuba-diving instructor or divemaster certificate, lots of diving experience
hrs/wk	varies
life–work	busiest times are weekends and holidays

The great white shark averages 3.5 to five metres long. The biggest great white shark on record was seven metres long, weighing about 3000 kilograms. As with most sharks, females are larger than males. Shark pups can be over 1.5 metres long at birth and around seven pups are born in each litter. These sharks take about nine years to reach maturity.

Careers in shark handling

There are many different careers that will allow you to work with sharks – in the tourism industry, the scientific world, in filmmaking and even in commercial fishing.

Sharks continue to both terrify and fascinate us. It's no wonder that the collective noun for a group of sharks is 'a shiver of sharks'.

> The great white shark is the ultimate nightmare for humans, nothing else on earth can eat us alive whole. They are the last great predators.
>
> David Doubilet, Underwater photographer, *National Geographic*

If it makes you feel any better, of the 360 or so species of sharks, only four are known to attack humans. In fact, while humans kill around 70 million sharks a year, experts estimate that only between 10 and 25 humans are killed by sharks each year – far less than the number of people killed by lightning.

> I believe implicitly that every young man in the world is fascinated by either sharks or dinosaurs.
>
> Peter Benchley, Author of *Jaws*

Early filmmakers and photographers tended to exploit people's fear of sharks. These days, though, people who work with sharks are often found fighting for their conservation, and are keen to dispel the myths that drive people to hunt and kill sharks.

While it is a small and highly specialised field, there are still opportunities for passionate shark-lovers to work with these creatures in the wild and in aquariums.

Here are some of the possibilities.

Aquarist: Oceanariums have become big business in recent years and there are opportunities to feed and care for sharks in an aquarium setting. Some aquariums require their staff to go and catch sharks from local areas.

Dive tour leader: Many dive companies offer their clients guided underwater tours of shark habitats in locations around the world.

Filmmaker and photographer: Although this is a very difficult career to make an income from – there is a lot of tough competition out there – there is still ongoing interest in films and images of sharks in the wild.

Marine or shark biologist: Biologists are employed in private or government laboratories researching sharks, investigating their environment, behaviour and physiology. Some teach or research in universities or museums. Most have at least some contact with sharks in their natural habitat.

Shark handlers are found in diverse places, but one of the more common industries is the tourism industry. Handlers are usually dive leaders who take their customers on guided dives to swim with sharks.

Do I have what it takes?

Taking the occasional dip can be pleasant enough, but it's quite a different thing when we're talking about a tank full of sharks – so don't dive in just yet! If you're willing to work in close quarters with these predators of the deep as a dive coordinator, you'll need a couple of things before you take the plunge:

- people skills – the sharks aren't the only potentially dangerous animal you're in the water with! This applies to your fellow dive operators and your clients, some of whom will have no diving experience whatsoever

- diving skills

- knowledge of shark behaviours

- respect for sharks – for their sake as well as yours. It helps to keep in mind that when you're in the water you're a guest in their environment.

Also, it might be a good idea to avoid watching *Jaws* for a while!

Who could I work for?

Depending on the area you specialise in, shark handlers find employment either in the tourism industry (as dive leaders or aquarium workers), in filmmaking (as underwater photographers and cinematographers) or in research (as scientists and marine biologists.)

Nick Kirby – Head curator, Melbourne Aquarium

Exhibition Aquariums are one of the biggest employers of shark handlers. Nick Kirby is the head curator of Victoria's largest aquarium. He told us about how he selects his staff.

What is your role at the Aquarium?

These days I do a lot of staff management although I still get the chance to get into a wetsuit pretty regularly. We don't have a high turnover here so it's not often that I am looking for new staff. I tend

to promote from within so most roles advertised to the public are entry-level.

We have literally hundreds of job applications whenever we advertise so there is a lot of competition, even for entry-level positions.

The sorts of things that my staff does – and I do too, to some extent – are the everyday maintenance of the exhibits. We feed the fish, clean out the big tanks and make sure that all the creatures are healthy. We also go out and catch fish – including sharks – for the aquarium.

And you get into the tanks with the sharks, right?

Yes, we do feed the sharks and get into the tank with them every day, as well as getting in the ocean with them – but there are occupational health and safety regulations that we follow very closely to make sure that there is very little danger. We never send a diver in with sharks by themselves and we always feed with a stick so that the sharks don't give you a nip. Our divers never dive alone.

What happens when you catch sharks in the wild?

We charter a boat and use a large tub and transport gear. We try to catch them locally – often we will capture a shark in Port Phillip Bay. We caught three last month.

We use small hooks and bait them carefully. When we catch a shark, we stretcher them over the side and remove the hooks gently. Actually they are not very violent or dangerous when they are caught – they are usually quite shocked. They will sit tight while they try to work out what is going on.

We have to catch animals that are a decent size – around two to three metres – so that they can go into the aquarium with the other sharks and not end up as lunch.

What sort of person would you look for if you were hiring someone?

We need people who have more than a marine science degree – that's the basic requirement, but you need more on top of that. We need someone who is an experienced scuba diver, which also requires a certain level of physical fitness. Ideally, holding a truck driver's licence is a good thing, because we need to be able to transport sharks and so on around the place.

You need to be comfortable and competent around boats. A boat licence would probably help, but we need someone who is able to go out and catch a shark – so quite a bit of experience in fishing is important. You can't do a course in fishing, though. That's just something that comes from experience.

What personal attributes help if you are working with sharks?

You must be a team player. You don't work with sharks on your own, so you need to be able to work well with other people in a team.

You need to be observant and to have a gut feeling for what's going on with animals. At the aquarium, we have a very strong focus on animal health so you would need to be able to quickly identify any potential problems.

Finally, you have to be passionate about sharks. There aren't many jobs around in this field so it is very competitive.

What sort of pay can people expect working in an aquarium?

That really depends on your location and your seniority. A small private aquarium in a regional location will pay far less than a big city aquarium with a dozen staff. I would estimate pay ranges from around $20K to $100K with most people earning $30K to $50K. It's not a job you do to get rich but there are plenty of other rewards.

What do you like best about working with sharks?

The two things that stand out for me are pride in creating a good environment for the animals – taking care of their health and so on – and the look on a child's face when they see a big shark close up for the first time.

Get started!

The best advantage you can give yourself in this field is to go diving and snorkelling regularly, and read up on all you can about sharks. Check out some of the documentaries on the Discovery Channel or hire them from a video shop near you – the more you know the better!

Another good way to learn about the marine environment is through fishing. Get yourself a fishing licence – you can fish from a wharf, from the beach, from a canoe or a boat – but most good fishermen reckon that to catch a fish, you need to learn to think like a fish. That's great training for someone wanting to work with sharks.

Owning a small at-home aquarium will give you a greater appreciation of maintaining marine habitats as well as the opportunity to watch the behaviour of marine creatures up close. Sharks are far from ideal candidates for a home aquarium but understanding the marine environment that they inhabit will help you understand them better.

Getting involved in scuba diving will also help you to make contacts in the underwater world. You can enrol in a scuba diving course from 14 years of age and you can snorkel at any age – you just need to be able to swim.

It may come in handy – and make you more employable – to have a boat licence, too. In most states, a restricted boat licence is available from the age of 12 and a general boat licence from the age of 16. You need to sit a test on marine regulations first.

The novel *Jaws* was inspired by a series of five fatal shark attacks that occurred in New Jersey over a 10-day period in 1916. Two days later, a great white shark was captured and blamed for the attacks. Experts now think it's likely that the shark responsible was in fact a bull shark.

If you are at high school or university, you may be lucky enough to find part time work in a dive shop. Often hanging around the shop and offering to work for free for a while will put you in the best position to be offered paid work when it does come up.

If you're interested in filmmaking or underwater photography it would be a good idea to invest in a video or stills camera. You can teach yourself the basics of filming and photography with practice and using some reference books borrowed from your local library. There are also many short courses you can take, often at local community colleges or TAFEs if you want to do photography.

Using a manual camera – for example, a 35mm SLR camera that has manual controls and interchangeable lenses – is the best way to learn the craft. You'll want to make sure that it's waterproof to a good depth, or that you invest in a secure underwater camera housing to suit your equipment.

Last but not least, you need to be confident around the water – non-swimmers need not apply! So get down to your local pool and start practising.

Courses and study options

Most professionals in the industry agree that a degree in marine biology will be a great advantage if you're interested in working with sharks.

Even though many current industry leaders don't have this qualification, there are far more people competing for a small number of paid work opportunities in marine work these days and many of them are already qualified marine biologists.

Most universities require you to have completed Year 12 maths, chemistry, English and biology to be eligible for entry. Some examples of undergraduate degree courses are shown below.

- Bachelor of Science (Marine Science), Sydney University

- Bachelor of Marine Science, James Cook University

- Bachelor of Animal Science, University of Melbourne

There aren't any TAFE courses in marine biology, but there are certificates in various other marine trades and industries which may show potential employers that you're keen to work around the water. Courses include marine services, marine mechanics and maritime operations.

FAST FACTS

The whale shark is believed to be the world's largest living fish, reported to reach lengths of 16 metres. The largest whale shark ever captured, off Baba Island in Pakistan, measured 12.6 metres in length and was estimated to weigh about 15 tonnes. However, whale sharks are not predators – they graze on plankton.

Sharks on the web

Underwater Australasia
www.underwater.com.au

Australian Society for Fish Biology
www.asfb.org.au

OzFishNet
www.amonline.net.au/fishes/ozfishnet/index.htm

The Shark Trust
www.sharktrust.org

Australian Museum's Ichthyology Site
www.amonline.net.au/fishes

So you wanna be a shark biologist...
www.elasmo-research.org/education/students/researcher.htm

Global Shark Attack File
www.sharkattackfile.net

Florida Program for Shark Research
www.flmnh.ufl.edu/fish/sharks/sharks.htm

The Great White Shark site
http://greatwhite.org

Crocodile farmer

'Never smile at a crocodile,' they say. But would *you* smile at a crocodile if you knew it was licking its scaly lips and getting ready to pounce at the sight of you?

Crocodiles are one of the last deadly animals still stalking the Australian wild – along with wombats. They kill dozens of unsuspecting swimmers worldwide every year, and there have been 13 known fatal crocodile attacks in Australia since 1985 (there have been no known fatal wombat attacks ... yet).

In all seriousness, the crocodile is one beast that was never meant to be contained. They are natural-born predators, and most crocodile farmers are quite aware that the larger crocodiles in their stock pens are watching carefully as their perfectly edible captors walk nearby.

 **FASTFACTS**

Crocodiles are the largest reptiles alive today. The larger species of crocodile can be dangerous to humans with the Saltwater and Nile crocodile killing many people each year throughout their range.

John Lever – Crocodile farmer

John Lever has been farming crocodiles since 1972. He is the owner and managing director of the Koorana Crocodile Farm in Queensland – home to over 3000 crocodiles. The farm has a tourist facility offering guided tours, a souvenir shop and restaurant. It also does its own skin and meat production and sells manufactured crocodile leather products.

John has four sons, all of whom have worked on the farm at some stage. His son Jason is currently managing the operational side of crocodile farming.

How did you become a crocodile farmer?

I was managing CSIRO agricultural research stations and was transferred to **PNG** in 1972. My role there was to establish a national crocodile farming industry, which we did over a period of about seven years. We had more than 200 commercial crocodile farms and tripled PNG's export income. Returning to Australia, I decided to establish a crocodile farm myself and set up in Coowonga, near Rockhampton.

What does a crocodile farmer actually do?

I make sure that we have the appropriate people in place so that the farm runs smoothly. My wife, Lillian, runs the tourism side of the farm, including the restaurant and retail outlet. I also have to market our farm production. Most of our skin products have been pre-sold for the next three years, but I am still busy making sure we have sales for the crocodile meat. These days, I am not as hands on with the crocs as I used to be. I am quite happy to stand back and let the young blokes do that side of it.

You had to start your farm by hunting crocodiles in the wild to get breeders – do you still do that?

No, it's not permitted in Queensland any more, although it is still legal in the Northern Territory and Western Australia. Nowadays we do captive breeding only.

We actually ran a crocodile removal service in Queensland for about 20 years. We would take them back to the farm for use as breeders. We probably removed over 100 crocodiles from the Eastern seaboard, in public recreational and residential areas. We would get called

when a crocodile was making a nuisance of itself – doing things like smashing chicken coops, eating the chickens or taking people's pet dogs. The largest one we picked up was about five metres long and weighed three quarters of a tonne.

Did crocodile hunting prepare you for a career in crocodile farming?

I actually learned most of the important stuff about crocodile behaviour in PNG. I worked with the crocodile clan people, who believed a creation story where a crocodile gave birth to their ancestors. They were great hunters and their skills were based around their knowledge of the crocodile's behaviour. I remember being told by one clansman, 'That crocodile is not ready to be caught' ... I didn't get it at the time, but after a while I started to understand what they meant. It is all about developing an understanding of how the crocodile thinks and being very, very patient.

Is it dangerous?

It can be, if you lose respect for the animals and get too confident in your own abilities. I've had a few tricky moments – I have ongoing back injuries thanks to a croc attack years ago in PNG. But our 26 years of farming crocodiles here have been injury-free.

How has crocodile farming changed since you started out in Papua New Guinea in 1972?

It's a well-established industry now, with defined markets, licensing and so on.

Better technology has certainly made crocodile farming much easier, more commercially viable and much safer. Pump-timer controls and electronic thermostats make a big difference and save a lot of labour.

The biggest improvement, safety-wise, has been the croc zapper. It's a portable electrical device invented by a bloke at Mareeba in conjunction with the Department of Primary Industries in Townsville. You have a pole with two prongs that that you put at the back of the croc's neck while it's in the water and it passes an electrical current through the croc's brain and stuns them for about four minutes. That's enough time to safely inspect the skin or shift them to another pen.

We used to loop a rope over their heads and drag them up. They would put up a good fight – you needed a lot of strength and a few helpers and it was always a bit risky.

What is involved in the production of crocodile skins?

When a crocodile reaches the right size, which is usually between 1.5 and two metres, they are processed in the abattoir and skinned. Nowadays we use an air-knife, which means that there is a lot less chance of putting a hole in a croc skin, which is important as they are worth about $600 each and you lose about half of that value if the skin is damaged.

Left to right: Crocodile ... and John

How do you monitor the crocodiles on the farm?

You do a daily check on the pens and just make sure that everything is okay. Our hatchling sheds are set up with closed-circuit TV cameras so we can monitor them from the office. We used to go and check on the pens in person and sometimes you'd have a killer – you'd find one or two dead crocs every morning. Or you'll have one croc getting much fatter and the others wasting away. It's quite common to get a bully who takes over the whole feed tray. The cameras help us to manage them better and isolate the troublemakers. And no, they don't know they are being watched!

What food do you give the crocs?

In PNG, we used to run into problems when farms would get bigger than the available food supply. They could really only grow if they were near an industry where they could use the waste product. Here, the chicken industry supplies the majority of crocodile food. Crocodiles eat chicken heads and the carcasses of old laying hens.

There's some research going on now to develop croc food in pellets. Making croc pellets will save a lot of work but there seems to be some sort of biological trigger missing to get them to start eating. Strangely enough, the crocs don't seem to like getting their food in little cubed pellets.

What do you like most about your job?

Crocodiles fascinate me. They are one of the oldest living animals on the planet – they go back about 180 million years. I really admire their capacity to survive. They also have their own personalities. They can be placid, bad-tempered – I can find a crocodile that has a personality similar to nearly every person I know.

What's the most important attribute for a future crocodile farmer?

It's really important to be alert and not get over-confident with the animals. You have to maintain a healthy respect for crocodiles to stay safe.

What sort of person is not suited to crocodile farming?

A lot of young men tend to get a bit cocky once they have learned the ropes. They can get a bit too confident in their own ability and take unnecessary risks. I have actually sacked all four of my sons at various stages for exactly that reason. It's not easy to maintain that level of care – women are much more sensible than guys as a rule.

fyi

John employs a number of people as farm hands, tour guides, process workers in the abattoir – and even cooking crocodile meat in the restaurant.

John and his son catch a crocodile.

I've been in a situation myself where I let my guard down for a moment and paid the price. Many years ago I roped a crocodile and tied its jaws together, and then I knelt down beside him to measure his head. He was a reasonable size, about three metres long. He knocked me down so hard with his head that I lost the use of my legs for three months. I was in traction in hospital in Papua New Guinea for a while and then they moved me to Brisbane where I had surgery on my back.

Do you need to be physically strong to handle crocodiles?

You should be fit and healthy but most of the crocodiles we work with are under 1.5 metres long so they aren't too heavy. The vast majority of crocs we work with are between 200mm and 1800mm. It takes them about three years to grow up to two metres, then we harvest and market them. The two metre ones weigh about 25 kilograms and they're very strong. If you're not careful they'll give you a nasty bite, which can bleed a lot.

You need a bit of strength to work with the big crocs, the breeders. I tend to stand back nowadays and let the young blokes do that sort of work, though it's much easier to handle them these days because of the improved equipment

What other personal skills are good for someone who works with crocodiles?

Patience is important – and the ability to understand the creatures. When you are working with baby crocs, you have to treat them as nervous animals as any unusual noise will frighten them. You also need a level of maturity. I wouldn't leave a young school-leaver alone near crocs, they take too many risks and put themselves in danger. Someone with a bit of experience elsewhere, working with animals in particular, might be better suited.

What training and education is important?

We train our own people, that way you know when they are ready to start working with the crocs. Some people will never be ready. A background in the sciences – biology or zoology – really helps.

in
brief

Crocodile farmer

$$$ 25–100K
age 63
quals animal handling and agricultural science qualifications
hrs/wk 40–80
life–work I live on the farm

Careers in crocodile farming

Farming crocodiles is not like farming cattle or chickens. One wrong move when dealing with a large breeder could be your last.

Yet the crocodile farming industry in Australia is a successful one. Crocodile skin is in great demand internationally for shoes and handbags and there is a small but growing market for crocodile meat.

Although there have been concerns from some sectors about whether it is ethical to farm crocodiles, farmers argue that a sustainable crocodile production industry is not only profitable, it also sets a controlled market for skins, helping to preserve crocodiles in the wild.

Crocodile farming is also far kinder to the local environment than farming cattle and sheep.

But, for crocodile farmers themselves, every day holds an element of danger. Farming crocodiles for meat is fine, as long as you remember that they might just be sizing you up for their own dinner plate.

> Crocodile predation on humans threatens the dualistic vision of human mastery of the planet in which we are predators but can never ourselves be prey. We may daily consume other animals in the billions, but we ourselves cannot imagine being meat for crocodiles.
>
> Val Plumwood, after surviving a 1985 crocodile attack, in *Being Prey*

Crocodiles are fast, aggressive and dangerous to humans, but they are fascinating to work with. At heart, they are wild animals and will never become a household pet. They can be contained – but never tamed.

It's not the kind of farming where you can relax around the stock at night.

Crocodile farming is a commercial business so there is a fair bit of administration involved for the manager and most farms need at least three or four staff to operate.

Some of the tasks involved in crocodile farming include:

- running a primary industry business – including the accounts, tax, administration, marketing and computer work as well as sales and exports
- feeding and watering crocodiles
- maintenance of pens, ponds and crocodile hatcheries
- making sure all of the crocodile enclosures are kept at the right temperature
- managing the breeding of large male and female crocodiles
- collecting eggs and baby crocodiles from the female breeder (very carefully!)
- managing egg hatcheries at the right temperature
- moving new and growing crocodiles between the various growing pens
- transporting crocodiles to an abattoir for processing
- selling skins and meat
- looking after other administrative matters such as government licensing and ensuring compliance with the industry regulations.

It takes two to four years to grow crocodiles from hatchling to market size.

Apart from a set of breeding crocodiles, which are fully grown adults, the majority of crocodiles on most crocodile farms grow up to about 1.5 metres before being killed and skinned.

Crocodiles can reach speeds of 11 km/h when they run, and even faster when they slide around in mud. They have been known to outrun a horse over a short distance.

Do I have what it takes?

Experienced crocodile farmers warn that this is not a job for risk takers or the faint-hearted. Come to that, this is not a job for most people!

For those who are up to it, there are a few things a good crocodile farmer needs:

- physical strength
- common sense – many young crocodile handlers have been badly injured and even killed when looking after their charges.
  Even old hands aren't safe all the time – in Queensland in 1994, an experienced crocodile farmer had his skull crushed by a crocodile that turned on him
- the ability to not get attached to the animals – a lot of them are destined to become fashion accessories!

At this point, we'll understand if you'd prefer to work with the kittens in your local pet store instead ...

Crocodiles are fierce predators and should be considered a serious threat to people. Although freshwater crocodiles are not usually known to attack, the saltwater crocodile is very aggressive and has been known to attack and kill swimmers.

Who could I work for?

Most large crocodile farms employ people to handle crocodiles, manage feeding and breeding, work in abattoirs or as scientists and vets.

Many crocodile farms also host tours and demonstrations and need people to act as guides and perform show feeding of crocodiles.

Most workers are employed either as tour guides or farm hands.

About 80 per cent of the income from crocodile farming comes from the skins, with a further 20 per cent from meat. Australia sells about 18 000 skins a year from saltwater crocodiles, which produce the finest quality skin of all crocodilian species.

The main markets for crocodile skins are in Japan, Singapore, France and Italy, while crocodile meat is sold in Japan, China, Europe and

Australia. Most farmers use an Australian agent who specialises in crocodile sales.

Although crocodile farms can theoretically be established anywhere the right temperature-controlled environment can be set up, in practice farming is done almost exclusively in the north of Australia, in Queensland, WA and the Northern Territory.

Get started!

Most crocodile farmers set up a farm from scratch and operate it as a business. It's an expensive investment, so it is wise to start by getting some experience in the industry by working on someone else's crocodile farm.

The Queensland Government's 2005 *Guidelines for Working with Crocodiles in Captivity* notes that although having an understanding of crocodile behaviour will assist in decreasing the risk of a crocodile attack, crocodiles are 'instinctive predators and can still be unpredictable animals'.

Crocodiles have eyes on top of their heads to help them spot any passing prey. They wait until the prey is close, then spring from the water to grab the prey.

Crocodiles do not chew their food. Instead, they use their strong jaws to rip the food into bite size pieces.

They recommend that anyone working near unrestrained crocodiles over 1.5 metres long should always have a colleague nearby to be on guard. It's their job to watch the crocodiles, tell their partner where they are moving and if necessary, distract the crocodile away.

Crocodile farms near you

Most crocodile farms in Australia are in Queensland and the Northern Territory, where conditions are best for breeding and catching crocs. Check out the websites of these crocodile farms around Australia – and be sure to make contact if you are ever in the area.

Melaleuca Crocodile Farm
www.users.bigpond.com/PWfisher/index.htm
Breeders and growers located west of Cairns, QLD; site has information about farming crocodiles, conservation, and research and development.

Edward River Crocodile Farm
www.pormpuraaw.qld.gov.au/crocfarm.htm
Western Cape York, QLD – Australia's first crocodile farm, established 1969. Run by the Pormpuraaw Aboriginal community.

Koorana Crocodile Farm
www.koorana.com.au
Commercial crocodile farm, tourist park and restaurant near Rockhampton, QLD. Site includes information on meat and skin products, tourism and farming, and educational material.

Hartley's Creek Crocodile Adventures
www.hartleyscreek.com

Near Port Douglas, QLD. Hartley's Crocodile Adventures is a tourist wildlife park and crocodile farming enterprise.
Site includes image library.

Johnstone River Crocodile Farm
www.crocfarm.com
Tourist site and crocodile farm, near Innisfail, QLD.

The Darwin Crocodile Farm
www.crocfarm.com.au
South of Darwin, livestock of 10 000 crocodiles. Site has information about skins, leather, flesh, prices and its history of crocodile farming.

Like all farms, crocodile farms have to sell their products at the price set by the international market – it's not really possible to determine your own price. The only way to increase your income is through efficient production and the sale of top quality skins.

There is some government support in the way of advice and tax breaks. The Queensland Government's Department of Primary Industries and Fisheries, for example, has a spreadsheet program called CrocProfit to help plan a crocodile farming business.

Saltwater crocodiles are one of the few animals that will eat humans readily as food. Most attacks on humans occur at dusk or at night, when the person is wading or swimming in shallow water.

In some states, crocodile farmers may be given a limited permit under some circumstances to obtain breeding animals from the wild. However, once established, most farmers grow their own breeding animals and then collect and hatch their eggs. (Even a ewe will fight like a lion to protect her lamb. Can you imagine what it'd be like taking eggs from a mother crocodile?)

A successful crocodile farm needs to be close to a crocodile processing plant, feed suppliers (such as a chicken processing plant), and transport hubs. Farms in remote sites often experience problems with high transport costs for crocodile food and for selling skins and meat.

Experts believe there are plenty of market opportunities for farmers producing high quality skins.

Processing crocodiles involves killing and skinning the animals and processing the meat. The skills used in skinning are quite specific and can have a dramatic effect on the quality of skin produced, so many crocodile farmers contract the work out to specialty processing plants, where there are also employment opportunities.

Courses and study options

At February 2006, there were around 112 registered courses across Australia that dealt with animal care. While some are university and postgraduate courses, others are entry-level courses offered by TAFE and private agricultural colleges.

Course qualifications range from Certificate level to Master's degrees, and everything in between.

Crocodile farming is a specialised field and having some training and experience working on someone else's farm for a period of time to thoroughly learn the ropes is highly advisable before breaking out on your own.

Animal handling courses are advisable, but if you are interested in establishing your own farm, some formal training in business management may also be very useful.

Choose from one of the many Certificate I to IV in Animal Studies or Captive Animals or Animal Technology and Diploma of Animal Technology courses.

Courses qualifying you to work with animals are offered by various TAFE and vocational training colleges throughout Australia.

More information:

NSW
www.newengland.tafensw.edu.au

Queensland
www.agriculturalcollege.qld.edu.au

Northern Territory
www.myfuture.edu.au

WA
http://psc.tafe.wa.edu.au

Victoria
www.nmit.vic.edu.au

South Australia
www.vlepub.sa.edu.au

Tasmania
www.opcet.tas.gov.au

www.petalia.com.au

Learn as much as you can about crocodiles. If you are still at school, it would be a good idea to study biology.

Practical subjects in the industrial technology field may also be useful.

You may even wish to get a reptile as a pet, although you should make sure that anyone else you share a house with is ok about that.

Laws about keeping reptilian pets vary from state to state within Australia. For example, in NSW you must get a licence from the National Parks and Wildlife Service to keep captive-bred reptiles as pets.

Crocodiles are obviously not household pets, but reptiles that can be legally kept as pets include snakes, lizards and turtles.

If you're still at school or university, a part-time job working with animals will look good on your résumé if you plan to apply for a job at a crocodile farm later on. Try working at your local pet store or volunteering at a nearby zoo.

Crocodiles on the web

Department of Primary Industries and Fisheries' work with the crocodile industry
www2.dpi.qld.gov.au/crocodiles/13465.html

Wildlife Management International
www.wmi.com.au

Rural Industries Research and Development Corporation
www.rirdc.gov.au

Living safely with crocodiles
www.mesa.edu.au/friends/croc_kit/default.asp

Crocodile specialist group
www.flmnh.ufl.edu/natsci/herpetology/crocs.htm

Australia Zoo – Home of the Crocodile Hunter
www.crocodilehunter.com

Animal Planet – Ferocious crocs
http://animal.discovery.com

Aerial musterer

Is flying your gig? Would you like to pilot a small plane in a remote area, working in really tight spots, herding stock from the air?

Aerial mustering is one of the deadliest jobs in the aviation industry – there's no margin for error when you're flying just 200 metres above the ground. You don't get second chances.

The topguns of the commercial flying world, these outback pilots coordinate hundreds of animals (such as cattle or sheep) from the air with the assistance of a ground crew on motorbikes. They fly in some of the most remote and harsh country in Australia.

It's a lonely life. The work is hard, the hours are long, the pay isn't that great – and, apart from all that, you have to be one hot pilot.

Reckon you've got what it takes?

Mustering, or 'rounding up' animals, is an essential part of raising stock. Cattle, sheep and goats need to be gathered up for all sorts of tasks such as shearing, branding and selling.

In the Australian outback, a single paddock can be tens of thousands of hectares. Mustering from the air saves months of work on the ground.

Mustering can be a huge exercise that spans a number of weeks and involves one person in the air and at least three people on the ground.

Ashley Dowden – Aerial musterer

 builder >> truck
driver >> pastoralist >>
pilot >> aerial musterer

Ashley has been an aerial musterer for more than 15 years. He lives on Challa Station, a sheep property that his family has owned for over a century and covers 87 000 hectares of semi-arid shrub lands outside Mount Magnet in WA.

Mount Magnet is around 600 kilometres north east of Perth on the Great Northern Highway. The surrounding pastoral area boasts some of the largest sheep stations in Australia.

Both Ashley and his wife Debbie are commercial pilots. They run the station along with their aerial mustering business, Murchison Aviation. The Dowden's four children attend the School of the Air.

Ashley grew up on Challa, trained as an auto-electrician and then became a **jackaroo**. After some time he obtained his pilot's licence and the rest, as they say, is aerial mustering. He is now the chief pilot for Murchison Aviation and a pastoralist.

 **gloss**ary

Jackaroo means:
– an apprentice at a station.

q&a

What is an aerial musterer?

A hundred years ago, a top stockman riding a nimble-footed animal was in charge of mustering. Now the aircraft has replaced the horse on stations throughout Australia, and the pilot has taken on the position of the stockman in mustering operations. The aerial musterer directs the ground crew to the stock and decides, in conjunction with the head stockman or station owner, the most efficient way to clear the paddock.

What does an aerial mustering pilot actually do, on a day-to-day basis?

The night before the muster, the pilot will usually discuss the tactics for the next day with the pastoralist. The pastoralist will get out a map of the station, point out the paddock and the watering points, and brief the pilot on the type of country to be covered.

The next day, the pilot will be airborne by first light, fly to the paddock and begin moving the stock from the point farthest from the exit gate.

What happens next?

The ground crew arrive – usually three to five people on motorbikes, horses or in vehicles. The pilot will direct each member to the mobs scattered throughout the paddock. The ground crew keeps the stock moving as the pilot directs them and searches for other stock in the paddock. Once the paddock has been cleared and all the stock are under the control of the ground crew, the pilot returns to base and the ground crew drive the stock in.

Is it a difficult job?

Absolutely. Of course, most jackaroos believe that the pilot has the easy job. On the ground, the musterers are often riding over really rough country, through cobwebs, getting flat tyres, getting lost and swearing and cursing at the pilot who can't understand why they can't see that mob of sheep on the other side of the mulga wash. And then the pilot gets home hours before the rest of them!

The pilot, however, has the enormous responsibility of the safety and efficiency of the muster, so it seems fair enough they be home first to get the best of the morning tea.

When did you start working in mustering?

I grew up on the land. Stock is in my blood. I'm an auto-electrician by trade but I can't imagine working off the land. I got my pilot's licence 15 years ago.

Why did you decide to get your pilot's licence?

It was a case of moving with the times. We have a property that's about a quarter of a million acres. The mustering was done on horseback in the old days, but these days ground mustering is mostly done by motorbike. With air support, a muster that might take two months can be completed in two weeks.

How often do you muster?

On our property, we are mostly farming sheep, so we move them for the lamb marking season in February. We move them again later for shearing – which we start quite early here, say, July. Then we sub-contract to other properties all over WA, bar the Kimberleys, for mustering.

What do you enjoy about it the most?

When you're flying, every day is different. You can work the same paddock three days in a row and the weather will make it different each time.

It's always exciting, always interesting. I'm my own boss and I have freedom and versatility.

There's also a high level of mateship in this job – you're working and living very closely with people for an intense stretch of time.

On the other hand I love the solitude of flying … you don't feel crowded up there.

Ashley loves a sunburnt country, a land of mustering planes...

Is it dangerous?

Flying at such low levels all the time is inherently dangerous. You are at 200 feet, usually flying fairly slowly and with not a lot of lift. If you stall you will be killed – there is no time to recover.

While you are flying you are constantly looking outside the aircraft and using the radio to guide blokes on horses and bikes below. Any lack of concentration could kill you.

It's a very fatiguing job with a heavy workload, and inexperience will make the stress levels even higher.

Having said that, it's only ever as dangerous as you let it be. We build in margins and teach these to people in our flying school. If an aircraft can fly at 50 knots, we don't ever go below 60 – that gives you a margin. Same with **banking**; the aircraft may be able to handle a 70 degree bank, but we will set your limit at 60 degrees.

How do you cope with the fatigue, personally?

I was born on the station, so you learn to work really long hours in the peak times – but when we sleep, we really sleep. I am in bed by about nine.

I aim to be airborne by 5.30 am. Then I'll head out and do about three to four hours of mustering, land on the airstrip at the north end of the property and drive the 4WD back to the sheep yards. I should be there just when the ground crew are turning up with the mob. Then I will work with them, processing the lambs. At the end of the day, we let the sheep out into a holding pen. Then I drive back to the aircraft and fly home at around 7.30 pm.

glossary

Banking means:
– turning an aircraft in flight.

Is there anything you don't like about this job?

For me, it's ideal. My wife Debbie is a pilot and we run the company together. This kind of life is very hard on most women though, and hard on the partners of the pilots. It is very isolated and most women don't like to live out in the middle of nowhere. Often the wives will run the financial side of a property as well. Most children do School of the Air.

How is the life–work balance in this job?

This is a job for single people. It's really hard to do it with a family. You are away for two to three weeks at a time, you leave before sunrise and get home in the dark and it is really tough for your partner and kids. It's a little bit easier for me because I run the business so I can have a bit more of a say about which jobs I go away on.

There are a lot of single blokes in their mid 30s out on stations who will probably never marry. It can be a difficult place for some women to live and it's hard for a young man to convince a woman to come out and live on a station, particularly when he won't be around all that much. In summer, it can get to 45 degrees. We have no air conditioning – we run on solar power here, with a backup generator, but most places you can't generate enough power, or justify it, to run an air conditioner. The heat and isolation can get to people.

in **brief**

Aerial musterer

$$$	5–50K
age	over 35
quals	pilot's licence, jackaroo experience
hrs/wk	60 including 20 hours flying time
life–work	always on the road

Careers in aerial mustering

If you are interested in commercial flying and you want a job where you can push yourself to the limit every day that you take to the air, you will love this job.

Low-level flying is extremely difficult and dangerous work. You have to manoeuvre a small fixed-wing aircraft (in the case of fixed-wing mustering) or a helicopter (for heli-mustering), and often do so in turbulent conditions.

Mustering pilots often have to be able to turn and bank at slow speeds and low levels.

It's just like acrobatic flying – except that it's always unrehearsed and often in unknown and unhospitable terrain – and you're also trying to manage 1000 cattle and four blokes on motorbikes at the same time.

Most of the time, you are required to work on remote cattle stations or sheep stations. It's a male-dominated world – women are outnumbered about five to one. You have to be prepared to rough it.

Mustering is a nomadic life. With little chance to establish a home, pilots move from station to station, job to job.

Commercial pilots have limits on the number of hours that they are permitted to fly, under Civil Aviation orders. However, mustering work usually starts just before dawn and ends just after sundown, so the days can be very long.

Since 1970, 10 people in WA have died as a result of aerial mustering accidents and it has been noted by the WA Coroner that young, untrained and inexperienced pilots are the most vulnerable.

Why on earth would you do it?

For some pilots, it's the opportunity to be the best.

Many agricultural pilots believe that their experience makes them better pilots than even the elite military fighter pilots in some of the world's best air forces.

For other pilots, it's the opportunity to combine their love of working on the land with their love of flying.

Still others relish the nomadic lifestyle and the opportunity to be involved in a unique occupation.

Finally, though your job is that of head stockman when you're supervising a mustering team, you're responsible for your own work. Aerial mustering pilots run their own show, and that independence is a big attraction for some.

I live as a winged nomad, flying between shearings or truck outs. I stay wherever my work takes me, for as long as they need me – and then move on. My home lies somewhere between Paynes Find Tavern and the Kununurra Racetrack. Or everywhere between.

Debbie Dowden, Operations manager, Murchison Aviation

Do I have what it takes?

Who knew looking after sheep would require a pilot's licence? The best aerial musterers have:

- an understanding of stock – this is probably the most important thing. You need to have quite a bit of ground mustering experience to be able to muster from the air.

- aircraft experience – you will need many hours of experience in the right type of plane before you are considered competent enough for this job

- an obsession with safety – if you take risks at the controls of a plane, you are not only jeopardising your own life, but the lives of others

- the ability to rough it if necessary. While the mustering pilot is usually put up in the family homestead, sometimes they may have to camp out – which may mean no hot showers or running water, cooking your own dinner and sleeping on the ground around a campfire with the rest of the ground crew.

Then, of course, there is the ability to get along with country people – you may have to supply your own akubra.

Who could I work for?

Aerial mustering companies are generally very small organisations employing two to five pilots. Because of the vagaries of rural life, a lot of companies don't stay in business for many years – long periods of drought can wreak havoc on this industry.

Flying schools are generally a good source of information about aerial mustering job opportunities in their local area.

The most common use of low level flying is for specialist agricultural flying such as crop-dusting and cattle mustering. Other uses include pipeline inspection, fire fighting, search and rescue operations and military operations.

Debbie Dowden – Operations manager, Murchison Aviation

Debbie Dowden is the operations manager of Murchison Aviation as well as a pilot. She is also a flying instructor and recently wrote the Aerial Mustering Code of Practice for Western Australia. She employs between two and five pilots.

What is your background and what is your role at Murchison Aviation?

I didn't grow up on the land but I was heavily into racehorses when I was younger. I did a Diploma of Education and also did my pilot's licence years ago. Then I worked at the Royal Aero Club of WA at Jandakot, training pilots, which is where I met my husband Ashley. These days, I look after all of the administration for the company. I take all the bookings for the work and make sure that the pilots know where they are going. I also keep track of the jobs, order fuel, book in the regular maintenance for the aircraft and that sort of thing.

What sort of person would be suited to work in aerial mustering?

Mustering pilots are a particular breed of pilot. They love the bush and are willing to face the hardships it offers. They are comfortable working with animals and they travel extensively to different stations and work stock alongside the station owner and stockmen. They are adaptable to all different types of flying and are alert to the dangers associated with low-level work. It might not be right for anyone who yearns for the glamour of a typical airline job – but it certainly is an exciting and challenging position for anyone with a sense of adventure.

What skills are important to make a good mustering pilot?

The single most important skill for a mustering pilot to have is stock sense. Things like how the animals will react to an aircraft, how they run under certain weather conditions, what type of country they'll graze on at a particular time of year and how they will move over that country. The pilot needs to know how to get the animals moving in the desired direction at the optimum speed. Almost anybody can be taught to fly an aircraft, but few pilots possess enough stock sense to run a muster from the air.

What makes a bad mustering pilot?

There is nothing more frustrating for the boss of the station to have an incompetent mustering pilot. They could spend more time lost than actually mustering stock. The ground crew have to take matches to light a fire so the pilot can find the paddock. Then when he does find them, he spends more time cutting stock off from the mob than driving them in the right direction because he simply does not understand how the animals will travel in a given situation.

What advice would you give to someone who would like to become a mustering pilot?

Stock sense is something that you develop over a long time working closely with stock. People say that you must do 10 years on the ground first – on a motorbike and under an aircraft – before your stock skills are good enough to enable you to be in charge of the operation. If you are seriously considering becoming a mustering pilot, your chances of employment will be greatly enhanced if you have at least had some experience with stock, preferably on a station.

Working on a station will also give you the chance to find out if you're suited to that type of life before you spend good money on your mustering rating. Station people can recognise a newcomer from Sydney and will go to great lengths to play practical jokes on anyone a bit wet behind the ears. They once convinced a young pilot that fence wires sagged because emus perch on them. There was another time that involved a small amount of explosives ...

On the other hand, station people are amongst the most honest and hardworking people in the country. If you can't get on with station people, you won't enjoy being a mustering pilot for long.

What training is important for a potential mustering pilot? Is there a difference between the minimum training and what is safe to do the job well?

A mustering pilot must have the appropriate qualifications before considering any type of aerial mustering work. In WA there have been three fatal mustering accidents over a recent 18-month period. Worksafe and **CASA** are extremely concerned about these fatalities and are considering changes to the aerial mustering industry as a result.

How did these accidents occur?

Most mustering accidents have occurred with inexperienced pilots, as a result of a stall at low altitude. The results are almost always fatal.

Some accidents have involved pilots who did not hold any type of mustering qualification. Perhaps they were pressured into or persuaded to take on the mustering because they held a pilot's licence. One wonders whether a few hours of good training could have saved their lives.

So, despite the reputation that mustering pilots may have as the 'cowboys' of the aviation industry, they must be safe. They shouldn't take risks. Mustering pilots need to be obsessive about safety, fuel management and adherence to the regulations. Too many people have lost their lives by ignoring these things.

If you are considering becoming an aerial musterer, it is vital that you hold a low level and mustering endorsement. The good news is that there's no need to have a **twin engine endorsement** or even a night rating, as mustering is done during the day and few stations have lights on their airstrip.

As a mustering pilot, you must be completely and utterly familiar with the aircraft you intend to fly, usually an older type Cessna 150 or 172 or a SuperCub. Our operation prefers to employ pilots with at least 500 hours of experience on a Cessna.

During a mustering operation, the aircraft becomes a tool to keep you airborne – control of it shouldn't require too much thought. All of your concentration will be on the stock and the ground crew. Your life can depend upon your ability to fly without having to think about it.

How difficult is it to get work?

We have had five years of drought here so there has been less than half the usual amount of work. Farmers can't afford to run stock and either they aren't mustering or they can't afford an air muster. A lot of the time we didn't have enough work for our own pilots. The only thing we could offer them was some fencing work or other work on the property.

10 years ago, we were flying an average of 2000 pilot hours each year. Last year, we ran about 400 hours.

However things are picking up so we are currently looking for another mustering pilot. The problem is that there are plenty of commercial pilots out there, but few with any experience with stock.

glossary

CASA means:
– Civil Aviation Safety Authority.

Twin engine endorsement means:
– an aviation rating qualifying a person to fly a twin engine aircraft.

I could employ a dozen pilots tomorrow with less than 200 hours flying time and no stock experience, but I don't want to kill them. Ideally we are looking for a young single guy or girl who has plenty of ground experience as a jackaroo or **jillaroo** or with stock, and who has enough flying hours to be safe.

What sort of wages can a pilot earn?

Pilots are paid an hourly rate, as contractors. Wages vary depending on experience and the number of hours that you are offered. You are only permitted to fly 1000 hours a year and the maximum rate is about $50 per hour, so you wouldn't be able to earn more than $50K a year, although I think that the average is a lot lower.

What are the working conditions like for pilots?

Most of the strips a mustering pilot operates from are reasonably well maintained gravel ones – but even these can pose a problem in the wrong conditions. Sometimes the pilot will be required to land the aircraft in a suitable area that is not actually marked as an airstrip. The pilot must be both confident and capable enough to adapt to any of a number of possible conditions. We have known some pilots who grow pale at the thought of **dangling the Dunlops** on 200 metres of dirt next to the shearing shed. They require 1000 metres of sealed runway and a windsock before they'll land an aeroplane. No stations offer this type of facility. Mustering pilots have to be prepared to land anywhere.

What about personal attributes?

The mustering pilot needs to be a sociable and flexible character. While on a job, he or she will be required to camp at the station, usually sharing the table with the boss and the crew as well as the wife and children.

Sometimes the mustering crew will be based at a remote 'outcamp' with few facilities and no luxuries. The pilot could be living with strangers and away from home and civilisation for weeks at a time.

We've been on jobs where there have been no showers. Most of the ground crew slept outside and shook the ice off their swags each morning. Other jobs are at premier tourist stations and offer really luxurious accommodation and meals. Most of the time, the pilot is comfortably accommodated in the family homestead.

What is it that attracts so many people to the job?

Being a mustering pilot is so much more than flying an aeroplane. It is a way of life with great possibilities and endless challenges. It is a wonderful way to see the country and meet characters as well as share the lives of the people in the bush. Apart from anything else, it's a chance to develop much sought after skills as an aviator.

Get started!

To become a mustering pilot, you should have experience in the agricultural industry – specifically experience handling stock. This is

glossary

Jillaroo means:

– a female jackaroo.

Dangling the Dunlops means:

– getting an aircraft ready to land.

essential training for a good mustering pilot. The best university in the world cannot teach you how to read sheep – the only way to do this is through experience.

Growing up on or familiarity with remote areas of Australia is a big advantage, but working with stock as a jackaroo or jillaroo on a station will give you plenty of useful experience. Try to get work on a property as a farmhand and work your way up.

Most pastoralists agree that it's easier to train a stockman how to fly than to train a pilot how to read stock.

There are no specific qualifications needed but if you're keen on study there are TAFE courses in animal care and agriculture that you can do.

You must have a commercial pilot's licence and a decent number of flying hours up your sleeve – so save up your money for flying lessons, and find out about flying schools near you.

Obtaining a pilot's licence is time consuming and expensive, but essential. Most employers prefer a pilot with over 500 hours of experience, ideally in the sort of plane that they will be using to muster.

There are many people holding pilot's licences who are looking for commercial agricultural work, but sensible pastoralists are reluctant to take on anyone who doesn't have experience with sheep or cattle.

So if you think you have what it takes to be a mustering pilot – muck in and get some experience!

Aerial mustering on the web

Aeroclub – Everything in aviation
www.aeroclub.com.au

Aerial Mustering Code of Practice for WA
www.pgaofwa.org.au/Pages/Committees/Pastoral/
AerialCodeEx.pdf

Leconfield Jackaroo and Jillaroo Training school
www.leconfieldjackaroo.com

Jumbuck Pastoral
www.jumbuckpastoral.com

Clearance diver

Have you ever watched a movie where someone has ten seconds to stop a bomb from going off? Imagine doing that – 60 metres underwater. For a clearance diver in the Royal Australian Navy, this is just part of the job – and there are no second takes.

In Australia, clearance divers work exclusively in the Navy. Clearance diving is one of the Navy's most adventurous and dangerous jobs, with only the fittest and most fearless candidates admitted. The work involves such a high degree of physical strength that it is one of the few jobs that females are not permitted to perform.

A day in the life of a clearance diver might include deep-water diving, exploration and dealing with dangerous weaponry, including mines, missiles, artillery shells and improvised bombs. The most important job for a clearance diver in the Navy is to find, identify and get rid of underwater military weaponry.

The divers need nerves of steel as they seek out and defuse underwater explosives. The lives of hundreds of their workmates can depend on their skill and knowledge. They can work day or night, on land or in water – sometimes at depths of 60 metres and beyond.

Over recent decades, there have been some combat situations where Navy personnel have played a key role. But more often, they're

involved in peacetime activities, like training and making sure they are ready for combat.

Some of the other jobs they might be required to do include underwater welding, repairs and maintenance of ships' underwater fittings.

Divers have to become experts in the use of small arms. They learn escape and evasion tactics, combat survival, abseiling and even parachuting.

CV athlete >> bosun's mate >> ship's diver >> navy clearance diver

Shane Smith – Navy clearance diver

Shane, now 29, joined the Navy 10 years ago. He is a clearance diver based at HMAS Penguin on Sydney Harbour and holds the rank of Leading Seaman. His work is extremely physically demanding and often takes him away from home for long stretches. Because he is involved in combat and counter-terrorism work, his name and certain personal details have been changed to protect his identity.

q&a

What do clearance divers do?

Basically we clear harbours so that it's safe for ships to come in. Our main task is to conduct diving operations as part of and to support the Royal Australian Navy, and to conduct exercises and operations with overseas forces. Mine clearance is the most important part of that. But there is also general diving and salvage work – clearance drivers have to be able to do everything. We are a very multi-skilled bunch.

Why did you become a clearance diver?

I love the ocean. I'm a fanatical surfer and grew up around the surfing industry. When I left school, I trained to be a sprint canoeist but couldn't get through to the professional ranks. I joined the Navy because I love the water and I wanted an outdoor lifestyle. I started as a bosun's mate. I was posted at HMAS Penguin, and there just happened to be a dive team based nearby, and I realised that I'd love to do that sort of work.

How did you get onto the team?

I applied to do the preliminary Navy ship's diver's course, which is a three-week **scuba** course that qualifies you to dive to 20 metres. Anyone in the Navy can do the course. I had never dived before, but I really enjoyed it. Then I applied to do the clearance diver's acceptance test, called the CDAT. It's a very tough, two-week course that is also a selection test for admission to the 27-week basic clearance diver's course.

What's the CDAT selection test like?

It's very competitive. You need to have the right attributes – not just physical fitness, but also mental toughness. It's a challenge just to be left standing at the end. Even if you get through the course, you still have to go up in front of a board of selectors, who then choose the people to go on to the basic clearance diver's course. When I did it, we started with 16 people and finished with eight – and only five were selected to go through to the next course.

What's involved in the basic clearance diver course?

The basic course is 27 weeks of training, very intensive – there are no breaks. You are sent to various dive sites all over the place, from Jervis Bay to Cairns. You learn maintenance skills, demolitions training and how to work with different equipment, so it's also intellectually taxing.

Is there any further training after that?

You start active service after the basic course, but as you move up the ranks a bit you're sent to other courses too. The intermediate clearance diver's course involves a number of modules that can take from six weeks to two months to complete. You might get 12 months to fit those modules in. Then there's the advanced clearance diver's course, which is 14 to 16 months long.

Once you're qualified, what sort of work do you do?

This job doesn't just involve diving. The main purpose of the dive is to find a mine. Once you've found it, you have to render it safe. That can involve a lot of land work as well – so it's not all water-based work. In combat we could be called on to do tactical explosive ordnance, which basically means disposing of weaponry. In the Gulf, the clearance guys were also driving around disposing of ordnance on the land.

As a clearance diver you're really a Jack-of-all-trades – there are so many different things you have to know. It's a very demanding job. You have to be constantly up to date – we're always learning to use new and very technical pieces of equipment. If there's anything you're unsure about, you read about it to find out what's going on.

Do you have to live on the base?

You can if you want to, but I don't myself as I'm married with a child. We live a few kilometres away.

gloss**ary**

Scuba means:

– self-contained underwater breathing apparatus. Divers carry a scuba tank on their back and breathe in through a regulator.

Clearance diving in the Navy actually requires some on-land work too.

Have you done a lot of travel in this job?

Yes, but I can't discuss the details. Most divers will do a number of trips overseas. The actual amount of travel varies. For about three years I was away eight months of each year, but usually I'm away a total of four months in a year. That might be one or two weeks at a time or it might be a longer stretch.

Is it difficult to find a life–work balance that suits you?

We travel a lot but we do spend some periods back at base, just doing the 7.30 to 3.30 routine. We need to carry out maintenance, do courses, take leave. In any given year you might get about seven weeks' leave. We sometimes work late at night or on weekends when we're at home. We have an on-call roster for weekends.

What's your daily routine like at the base?

We always start with physical training usually followed by some diving. Some days involve a bit of paperwork. Dive equipment involves lots of maintenance and we have to maintain it at a very high standard. Training might involve some bush navigation, patrolling, shooting and boat work.

What parts of your job do you enjoy the most?

I like jumping out of helicopters. I love cruising up Sydney Harbour in my wetsuit, knowing that everybody else is at work in an office. The most exciting thing I've done was fast-roping onto a high rise in the middle of Sydney. One side was 20 feet off the ground, the other side was about 600 feet.

Is there anything you don't like about your job?

Sharks, crocs and icy cold water. As long as a diver is warm and has good air, it's fine, but it can be miserable when you're cold.

I had a job in Darwin recently and just before we got into the water they told us there had been a big crocodile sighted earlier. But the worst time was doing some deep diving off Port Lincoln, where there were a few great white sharks around. I spent a lot of time peering over my shoulder. We do have devices that repel sharks, but we don't use them much – we're tougher than that!

Detonating explosives underwater is all part of the job.

Have you ever been in a dangerous situation with explosives involved?

Explosives are quite safe if they are handled correctly. The scary situations are when you're worried that somebody is going to get hurt. I had somebody pass out underwater once and we had to bring him out – those are the worst moments, when it's one of your mates. We always try look out for one another. We are a pretty close-knit community.

Are you an adrenaline junkie?

I have my fears, like anyone else. Having good training and good equipment takes some of the fear out of it. I do like the rush of doing some things.

What makes a good clearance diver?

Someone who is highly motivated and driven. You've got to be physically and mentally fit. You've also got to have the ability to think for yourself and a lot of trust as well – if I tell a diver to go and do a job, I can't see what he's doing under the water, so I've got to be confident that he can do it right.

What advice would you give someone who's after a career as a clearance diver?

You could start by jumping on the Navy website to find out as much information as you can – be prepared and be motivated. I believe you can actually join the Navy as a diver these days. A lot of the attributes that apply to a professional athlete might apply to a clearance diver.

in brief

Navy clearance diver

$$$	classified
age	29
quals	Navy ship's diver's course, clearance diver's acceptance test, basic–advanced clearance diving courses
hrs/wk	40 on base
life–work	varies

FAST FACTS

The maximum recommended depth for scuba diving is usually 40 metres. Trained Navy divers regularly dive at depths of 60 metres and beyond. It's rumoured that they can go to depths up to 90 metres, but the Navy won't confirm this for strategic reasons.

Careers in clearance diving

Diving with sharks and crocs, blowing up mines, jumping out of helicopters. Exciting work, yes – but also very exacting. One mistake could cost you your life, and the lives of those around you. But the adrenaline rush is one of the reasons people do it.

> Being part of an elite team in the clearance diving branch is one of the best jobs in the world. We start the day with a run around Sydney Harbour, over the bridge, to the Opera house and back. Where else could you get paid the sort of money we get to spend 90 minutes running around the foreshores of Sydney keeping ourselves fit? It's a great job.

Lieutenant Chris Reece, Executive officer, Clearance Diving Team AUSCDT One, Royal Australian Navy

It may not seem like there are too many job opportunities in this industry, but don't forget – there are more options that just being a Navy clearance diver. If this is the kind of career that interests you, think about what it is that appeals to you most about it. If it's the underwater adventure aspect, you could look into other careers involving deep-sea diving, such as working for an adventure company. If it's the weapons and bombs side of things, there are dozens of other jobs in the Navy and army that involved this sort of activity.

Scuba tanks usually contain compressed air, which is about 80 per cent nitrogen. At great depths, normal air can result in nitrogen narcosis (a dangerous intoxication) or decompression sickness – bubbles of nitrogen in the bloodstream. Both can be fatal, so a special mixture of gas is used beyond 40 metres.

Do I have what it takes?

To make it to the Navy's elite clearance diving team, you need to have some special attributes.

A very high level of physical fitness is crucial at any level in the Navy. But the best divers have other qualities that set them apart, such as:

- a 'never say die' philosophy
- high levels of perseverance and concentration
- reliability
- ability to work in a team
- ability to keep a cool head under pressure
- psychological stability and mental fitness
- attention to detail
- an easygoing personality – you must be able to work with a group of people at close quarters for long periods of time
- intelligence
- leadership qualities – especially to progress to management, you need to be able to manage staff and plan dangerous operations to the finest detail.

It might also be handy to remember that breaking into rousing choruses of *In the Navy* may be frowned upon.

Who could I work for?

There is only one employer for clearance divers – that's the Royal Australian Navy – so if you're keen on this job, you'll have to join up.

There are 13 000 permanent full-time staff in the Royal Australian Navy. Of these, only about 200 are clearance divers. Most are stationed either in Sydney, Perth or Cairns.

You can join the Navy either as an officer (requiring a university degree) or a sailor. Officers sponsored through the undergraduate scheme are required to complete a Return of Service Obligation – which is the length of your sponsorship plus one year. Those joining as a sailor usually enlist for between four and six years.

After your initial serving period has elapsed, you're free to leave the Navy or can choose to sign up for another fixed service length.

The first dive suit was developed in 1828 – a metal helmet allowing a diver to breathe underwater. The main problem was that the diver couldn't bend over without flooding the helmet!

www.defencejobs.gov.au

Chris Reece – Executive officer, clearance diving team

Chris Reece is a serving Navy officer who heads an elite clearance diving team. He's also involved in staff selection, management and training.

What do you look for when selecting a clearance diver?

The ship's diver's course is three weeks but it's not all that arduous. We need something to decide whether somebody has the aptitude to be a clearance diver, which is what **CDAT** is designed for. It's a two-week test with long hours, late nights and very early mornings. People are in and out of the water at all hours, so they are cold, wet, hungry and homesick. We have a psych officer on hand to assess the psychological suitability of the candidates.

glossary

CDAT means:

– Clearance Divers Acceptance Test

We're looking for people who basically don't stop, who just keep going through any adversity. They have to be very resilient, mentally and physically fit, and very good team players. The clearance diving team is a relatively small unit and we have to be able to work together very well. You can't have an individual who sorts out their own gear and doesn't help others.

What's the failure rate?

It's about 40 per cent these days – the CDAT weeds out people who don't have the physical skills or the aptitude to go through the long course. Diving is something where certain illnesses and injuries will preclude you. One bloke had two weeks to go on the course, then had a motorbike accident, broke some ribs and punctured a lung. We had to drop him. You can't dive with scar tissue on your lung – he was permanently medically unfit to dive.

What sort of person makes a good diver?

The teamwork aspect is the most important. Not everyone who starts the course is the fittest bloke in the world, but the ability to keep persevering with whatever you're doing and a 'never say die' attitude will get you through. It's not just the physical side – it requires excellent mental agility as well. Joining as an officer means they have to handle the bomb disposal side of things as well, so you need the ability to manage a steep learning curve.

What tips would you give someone considering clearance diving?

If you're thinking of applying, you need to keep yourself very physically fit. You should be fit on the inside as well. I know it sounds corny, but it will help if you scrap the junk food and stick to a really healthy diet. You need to show motivation and commitment in whatever you are doing. A level of teamwork is essential – we really emphasise the teamwork aspect. Being involved in sports, particularly team sports, will help you a lot.

The next step is to go to our recruiting office and talk to the advisers there.

How dangerous is Navy clearance diving compared to other Navy jobs?

We get paid extra 'danger' allowances beyond what others get, so it is recognised as fairly dangerous. In wartime, certainly, it is one of the most dangerous occupations. Diving with live mines in zero visibility at night, trying to fumble around and make them safe is certainly not easy.

Even during peacetime we play with explosives and go shooting on a regular basis. We're always diving to depths of 50m and beyond, which is inherently dangerous. There are certainly a lot of things that can go wrong.

However, the Navy has an excellent safety record in terms of our diving personnel. The last peacetime casualty we had was in 1983, and that was a crew member involved in a car accident.

Is it worth it?

Absolutely. Personally, I think this is the greatest career in the Navy.

Get started!

To join up you should be an Australian citizen, aged 17 or over. You will need to pass the Navy's stringent security clearance and be medically and physically fit for entry.

All Navy applicants have to pass a physical fitness test of 15 military push-ups, 15 military sit-ups and a **shuttle run**. Clearance divers do a similar test, except with 45 sit-ups.

Navy entrance training involves a four-week basic seamanship course followed by a month of small arms training. The ship's diver's course is a three week course and the CDAT goes for a very demanding two weeks.

By the time you have completed the basic clearance diver's course, you will have to be incredibly fit. Navy divers are required to run four kilometres in under nine minutes, then do 50 push-ups (in time with a metronome), followed by 18 chin-ups and 120 sit-ups. Then they

find out more

For more info on the Navy and how to sign up, check out www.navy.gov.au

Get trustworthy information about eating well at

www.nutritionaustralia.org

glossary

Shuttle run means:

– a demanding series of 20-metre sprints.

must swim 500 metres on their backs with fins (no arms) in under nine minutes and 15 seconds.

It's an unusual applicant who can match that level of fitness when they first apply – but you are expected to be very fit.

Having a non-Navy scuba qualification doesn't make much difference. The Navy has exacting standards and will retrain you through their course anyway.

Check the Defence website to find out as much as you can about clearance diving with the Navy.

> If you've got a general love of the outdoors, and you're physically fit and motivated, you'll love this job. The main thing is to have that 'never say die' attitude – like a lot of professional athletes. Team players do best here. It's very much a boy's world – it's a bit like being on a footy trip all year round.
> Shane Smith, Navy clearance diver, Royal Australian Navy

Preparing for one of the Navy's most physically demanding jobs is a big task. Not only must you maintain a high level of physical fitness but you also need to be prepared for the mental challenges you will face.

> Playing team sports at a high level and keeping up an excellent fitness level will help you to get ready for this career.
> Lieutenant Chris Reece, Executive officer, Clearance Diving Team AUSCDT One, Royal Australian Navy

FAST FACTS

A growing sport for divers is extreme underwater ironing. Participants take an ironing board and iron and set it up on the bottom of the ocean, and photograph it. Australia set the world record for underwater ironing with 70 underwater ironers in a pool in Victoria. (Although underwater ironing is unlikely to impress the Navy, you will be expected to maintain an immaculate uniform.)

Diving on the web

Dive Oz
www.diveoz.com.au

Scuba Diving Australia
www.divinginaustralia.com.au

Undersea Explorer
www.undersea.com.au

Sydney Dive Academy
www.sydneydive.com.au

Aquatic Adventures
www.acquaticadventures.com.au

Diving Careers Australia
www.divingcareers.com

Underground blaster

There's nothing quite like a really, really big explosion – that deeply satisfying BOOM, the smoke, the dust, the rumble of the earth beneath you ... Well, believe it or not, there are people in Australia blowing stuff up on a daily basis, and they're actually getting paid for it.

Welcome to the world of the underground mining blast crew – one of the few jobs where you'll be paid a lot of money to destroy stuff.

The mining industry is a rough and tumble world. Giant trucks haul tonnes of materials around all day, while far below the ground an enormous labyrinth of tunnels is filled with hundreds of miners drilling through kilometre-thick rock. Occasionally, the grinding of machinery is punctuated by muffled explosions from deep underground ... This is the blast crew at work. They are the toughest of the tough.

Often called the charge-up crew, they are the ones who explode rock by the metre to create the huge underground tunnels used for metal mining.

There are two areas where underground blast crews work – development and production. The blast operators (the 'charge-up operators') are responsible for setting up explosive charges to break or dislodge rock faces.

Like most jobs in mining, there is a lot of responsibility involved.

After a drill crew has bored holes into the site where explosives will be used, the charge-up crew will inspect the blast area and make sure all of the safety requirements are in place.

Next, they measure the correct quantities of explosives and put the charges and detonators into place in the rock holes. The blasting circuit is connected and tested – and then all miners retreat out of the mine site for the explosion, which is detonated remotely.

Once the charges have been fired, the charge-up crew must inspect the area to ensure that all of the charges have detonated and that loose rock hazards, mine roof supports and so on are controlled before declaring the site safe.

CV uni student >> bar worker >> fruit picker >> call centre operator >> meatpacker >> miner >> underground blast operator

Bryce Adams – Blast operator

Bryce is 22 and works on a large copper mine in north west Queensland.

Unlike many of his underground colleagues, Bryce has a degree in mining engineering. But, like all of them, he has worked his way up through the ranks of what he calls 'the underground hierarchy' to work on the charge-up crew.

Bryce went to university straight from school and studied engineering. His first full-time job was with a mining company and he is keen to continue his work in the industry.

When did you start working in mining?

I left school in 2000 and studied mining engineering at the University of Queensland. I graduated last November. After third year, I went to WA in the uni break and worked as a casual labourer in a nickel mine, operating machinery.

Why did you decide to work in mining engineering?

The first year of engineering is common for all strands of engineering so you have a year to make up your mind. Mining had the dollars and the job opportunities and sounded exciting.

Is the pay really that good?

At the moment, I'm on about $400 per shift. I go up to the mine and do 14 shifts in a row, then they fly me back to Brisbane and I get one full week off. When you're at the camp you don't spend any money, all your food and accommodation is supplied, there are volleyball and basketball courts and a pool – it's like a resort.

How remote is it?

We are 200 kilometres north west of Mount Isa and spend most of our time about 900 metres underground.

Has it lived up to your expectations?

Yes, definitely. I ended up in this massive industry with massive machinery … and massive money. On a normal day, your job might be to fix a piece of machinery that is worth over two million dollars.

How many of the people you work with have an engineering background?

Most of them are trained on the job. They might get a job through somebody they know or have experience elsewhere that gets them employed.

What experience do you need to have to work with the charge-up crew?

There is an underground hierarchy that you have to work your way through. You've got to be lucky to get a start in the mines. A lot of people start with driving a truck – that's probably the bottom rung. Next up would be a nipper – you run around the mine, getting things for everyone and making sure that everyone else has got everything they need to get the job done.

Next up is the service crew, which is in charge of extending all of the services underground, such as ventilation, lights, compressed air, water lines.

From there, you've got charge-up development and then, just above that, charge-up production. I'm going between both of those roles at the moment.

What does your job involve?

In development, you charge up all the development faces and stones – so that's the flat planes, tunnels and so on when you are advancing. You have a little charge-up wagon called a normet. You get the whole rock face and bore it with holes. Then you go and fill each of the holes with explosives, and you put little detonators in them. You might have 50 or 60 holes in one face and you have to make sure they all go off in the right order.

What's the difference between charge-up development and production?

Once the development crews have come in and charged the whole level, the production crew will come in. Instead of charging holes in the face in front of you, you charge holes in the roof above you. You are charging big rings at a time, like a big fan. Then you need to retreat back along the drives.

I like production more than development.

What do you like about it?

Well, you've got to be really careful. It's a very dangerous underground mining job. You have measures in place to make it safe, but you need to be able to make that call – is it definitely safe enough? You need to be able to assess that, and if it's not you have to tell the shift boss to stop the charges. There is a lot of responsibility in that role.

If you decide to do something stupid, like charge when it is not safe, and something goes wrong, then it is entirely your fault.

Is it life threatening?

Yes, definitely. The dangerous part of it is that the rings above you might have been retreated back 50 metres. The next level might be 40 metres up, so you will have levels on top of levels and there will be a massive void above you, deep underground. Rocks can fall down, **stopes** can collapse. And you've got to go right to the edge when you are in charge-up production, because you take the next ring off. You are right on the edge.

glossary

Stope means:

– a void created in the mining process.

106

Are you dealing with a lot of explosives?

We can blow up probably two tonnes of explosives in a day. The holes are drilled into the rock and we load the holes up with explosives. We hook them up with a remote detonating device and go up to the surface to detonate remotely. You couldn't be there when you blow it up – you wouldn't stand a chance.

Are the explosives volatile?

They used to be, years ago, but they are pretty safe these days. All of the bulk explosives are pretty insensitive – you can throw them around, walk on them, whatever, and they won't go off. They have to be set off with lots of little charges before they will go off.

How do you get the training you need to do this work?

You can do courses outside but most people get on-the-job training – the company organises the courses for you. I was trained by a guy who dropped out of school after Year 9, but he knew it all very well.

All of the training for your qualifications and tickets are provided, but there is certainly paperwork you have to do for it – things you need to learn and procedures you have to follow. You complete the theory training, which covers everything that you need to know, and then you have a training permit. You're then hooked up with someone who's worked for many years and has a wealth of experience, and they will supervise you for a certain number of hours. Then there is an assessor who comes down and assesses you. If you pass, you are then qualified to do that job. Training is an ongoing thing. You need to prove yourself at each level in this mining hierarchy before you can go any further.

What personal traits do people working in this sort of environment need?

You need to have skill and you need to put in the hard work and be prepared to give it a go. The main thing is your attitude – don't whinge, just get the job done. You definitely need to be a really hard worker.

How do you find the life–work balance in this job? Is it hard to catch up with friends and family and maintain a relationship?

I work 14 days straight and then have a week off. At the moment, it's awesome. I love it. I don't spend a cent for two weeks, because while you are on site everything is provided for you – accommodation, food, volleyball courts, basketball courts, pool, river – it's like a resort. You come back, you're loaded up with cash, ready to go and just party. It's harder for older people though, if they have families, not seeing them. There are plenty of towns where you can live at the mine site, though.

What do you like best about the job?

I love the roster – the money, conditions, the hours of work and so on. In the work itself, I have a lot of responsibility to make the right

decision and I like that. I work for 12 hours each shift but every day is different. Some days I might have four stopes to charge – in that case, it is going to be busy, I am going to be on my feet all day. I am going to have to move people all around me so that I can get it done. But the next day, I might only have two jobs. Then, I will often go and 'offside' for someone else, in the job above me, to learn what they are doing.

What jobs are next up the mining hierarchy?

After charge-up, the next role is the bogger or loader. It's a very big piece of gear – a bit like a front-end loader designed for underground use, because it's got a lower profile and can go through small tunnels. Once the charge has fired the rings off, you then go in and dig them out with the machines, load the trucks and take them away to the stockpile. This thing has a 17-tonne payload bucket on it.

Next job up is the drill rig. One is called solo and uses a bigger drill, which goes through and basically drills holes into the roof. The other is the jumbo, which requires a bit more skill. You go into new areas and drill the faces out – you might have 55 holes to drill. You have to make sure they are all the right shape and so on, so there is a lot of precision involved. You have to do all of your tickets and procedures to be able to work in that area, but you learn a lot and get experience if you are off-siding with someone.

Where do you see yourself going next?

I want to stay underground for a few more years. Engineers have to spend at least nine months underground, but I took this job because they agreed to let me stay underground for three years. I enjoy it and I also think having more experience underground earns you more respect from the miners.

Are there many women working in the underground mines?

Yes, there are. The women usually work together on the same crew. My sister is currently doing a mining engineering degree. The companies are trying to get more women into the industry so it's a good time to try to get in.

Would you recommend to this job to anyone?

It is a great job for a young person. A lot of people are discouraged from going into this sort of work because of the remoteness, but I don't find it a problem. They fly you in and out. While you are there you don't want any other distractions, you just work hard and sleep. Then you fly back to civilisation and you've got a whole week.

 **FAST FACTS**

There are a range of jobs that require a shot-firer's licence, which is awarded after the completion of basic training in the use of explosives and a satisfactory police check.

Careers in underground blasting

If you are into blowing stuff up and you want a job where you can
handle explosives as well as supervise blasting – you will love this job.

You can assure your mum that underground blasting is not mining's
most dangerous role; the most dangerous mining work by far is
underground coalmining.

No explosives are permitted in underground coalmining. In fact,
anything that could create a spark, like matches, cameras, even a
watch – are forbidden in underground coalmines.

FAST FACTS

Explosives are used in underground and open-cut mining,
civil engineering and for a range of other commercial
purposes including rock breaking, tree stump removal, pest
control, construction and demolition.

The temperature in the mine varies according to depth, but
a regular mine down to about a thousand feet will hover at
around 16 degrees Celsius year round.

Although metalliferous mines are not nearly as hazardous as
coalmines and there are many safety precautions built into the role of
underground explosives operator, it is still quite dangerous.

Be warned – you will also spend also lot of time underground in close
quarters with other miners.

The hours are long and the work is hard – but the pay is fantastic.

> I really enjoy my work. It's definitely the most dangerous job
> in the mine. You have to be very careful. You have massive
> responsibility and you make decisions about explosions that are
> pretty extreme.
> Bryce Adams, Charge-up crew, Byrnecut Mining

There are dozens of different types of jobs in the mining industry. The
Minerals Council of Australia estimates that the industry directly and
indirectly employs around 301 000 Australians. However, only a small
number of these are involved in using explosives.

Have I got what it takes?

It may come as a surprise, but mining crews don't let just anyone off the street get at their equipment. A good underground blaster needs to have:

- solid organisation skills
- the ability to work hard for long hours – some shifts last for 12 to 14 hours
- excellent attention to detail
- mechanical skills – some of the equipment is quite sensitive
- the ability to work in a harsh environment – the belly of a mine is certainly no place for claustrophobics
- great physical stamina.

There's a lot more to this job than just enjoying blowing stuff up!

detour

Find out about other jobs that use explosives – check out the pyrotechnist role in Career FAQs *Weird and Wonderful*

www.careerfaqs.com.au

Who could I work for?

The mining industry is huge and undergoes constant change.

Australia boasts hundreds of mining companies of various sizes and many miners are employed through sub-contracting companies, so there are loads of potential employers.

The best way to find an employer is to check a current list of mining companies for contact details. Organisations such as the Minerals Council of Australia list potential employers on their website.

Staff for mining companies are recruited in a number of ways. The selection of staff can be handled by managers working for a large mining company, owners and managers of smaller sub-contracting operations or by specialist recruitments agencies.

FASTFACTS

New mining technology and increased awareness of health and safety issues means mining accidents are on the decrease. In the 1940s there were 75 mining deaths per year. By the 1970s this had fallen to 25, and now there are only about 10 deaths per year.

Steve Hoggett – Recruitment agent

Steve has many years experience in human resources and recruitment for the mining industry. He spoke to us about what employers are looking for – and the best way to get that job.

What advice would you give to someone who wanted to work in mining?

The first thing is that you have to make a decision on what kind of mining you want to do. You can work in a metalliferous mine or a coal mine – and you can work in **open cut** or **underground mining**.

Each area requires very different skills. Some skills will transfer across – for example, if you have worked in charge-up underground, you probably have a shot-firer's ticket, which is used in both underground and open cut metalliferous mines and in open cut coalmining. Once you have a shot firer's ticket, it is just experience in the specific types of blasts that you need to gain.

Is it useful to get a shot-firer's ticket before applying?

Pretty much anyone can get a shot firer's ticket – subject to a satisfactory police check – but getting that ticket does not guarantee you work. It would probably be better to gain some experience working as an offsider in a charge-up crew and learn the skills. A shot firer's ticket issued by training organisation does not give you the experience that you need to be able to do the many varied kinds of blasts that are required in the mining industry. You would be better off gaining experience.

There are a number of explosives jobs you can do that can give you transferable skills – for example, working in civil engineering. You can probably get a start more easily in the civil area than in a mine. The industry is less well paid than mining, but explosives operators in those jobs need to work in quite close quarters with buildings.

Explosives operators have to be very accurate, using very small charges to move something slightly, rather than just blow the hell out of it. It's the quarrying and the civil blasts where the highest level of skill is required.

How difficult is it for someone to get a job in mining which will lead to charge-up work?

There's a lot of competition for places. I recently placed an ad for 20 operators with a large mining company and had about 2000 applicants.

Of those 2000 people who had applied for that job, half of them had tickets. The tickets themselves don't count for as much as experience. Only 500 of them had any experience. Those 500 were looking to move perhaps because this was closer to home or because they

Open cut mining means:
– a form of surface mining. Minerals are extracted from the earth through a shallow open pit. Sometimes called open-pit mining or open-cast mining.

Underground mining means:
– excavation that requires tunnelling under the earth. Also known as sub-surface mining.

worked for a smaller contractor – so someone with no experience obviously has little chance against those odds.

That example is for coalmining work on the east coast. But if you are fair dinkum about working in the mining industry, there are more mines and less people in WA, so it's easier to get a role operating in the west that it is in the east. Our west coast office is always asking us for any candidates we can bring forth.

What about the east coast?

Here we're sometimes able to take on somebody who has no more than an **HR** truck licence and some kind of mechanical background, starting on $80 000 per annum – but it's skilled people who are in demand and competition for spots is tough.

Qualified experienced tradesmen who work on even time rosters will be able to earn in excess of $120 000 – that's normal. We look for electrical trades, fitters, boilermakers – people who have industrial experience that can transfer to mining.

The more skilled you are, the easier it is.

What are the downsides to working in the mining industry?

Mining is well paid – but there are trade-offs for that. So many people go into mining and they have been there for 10 years – you can grab them by their ankles and shake them, and there's nothing coming out of their pockets. The lifestyle can lead to broken marriages and instability. A recent newspaper article pointed out that a Mackay Holden dealer was the largest regional seller of HSV vehicles. That is just one example of how the money gets wasted.

glossary

HR means:

– Heavy Rigid licence. Holders are eligible to drive a vehicle with three or more axles, for example a bus or a truck.

FASTFACTS

A little known snippet of trivia about explosives is that they can be used to tenderise meat. A tough old slab of steak can be tenderised by putting it in a tank of water and exploding a small amount of explosives in there. Don't try this one at home!

Get started!

There's no point just saying that you want to get into mining. You need to work out what it is that you want to do in mining, and then you need to get the skills or the experience in other industries that will make you attractive to a mining employer.

Courses and study options

A background in mechanics or a trade will be very useful in getting your blasting underway, so it might be an idea to do a TAFE course to brush up on your knowledge. Some of the options are listed below, most of them available from certificate level right up to an advanced diploma.

- Electrical engineering and wiring

- Mechanical and manufacturing engineering

- Explosives handling, storage and use

- Extractive industries operations

- Metalliferous mining operations

You don't need a university degree to get the skills you need as an underground blaster, but a Bachelor's degree in engineering might get you on the fast track to the mine itself. Here are some examples of university courses that might be of use.

- Bachelor of Engineering (Mining Engineering), University of New South Wales

- Bachelor of Science (Mining), Curtin University

- Bachelor of Engineering (Civil and Structural Engineering), University of Adelaide

Do some research on the Internet and have a look at what projects are on. Start with the blue-chip miners and work your way down to the contractors.

Apply for anything that is going with the main mining companies. Contact the recruitment staff and find out what they are looking for. Do any work experience you can lay your hands on. Don't just rely on school work experience.

A lot of people wait for somebody to do something for them. Don't. Go and work for free somewhere to get some experience, because it's unlikely you'll get that opportunity on a mine site.

There are opportunities in civil companies in the cities, or even in electrical companies. Try the work out and see if you like it!

Blasting on the web

Mineral's Council of Australia
www.minerals.org.au

Mining Australia
www.mining-australia.com

Careermine – Mining jobs and employment opportunities
www.infomine.com/careers

Mineral Policy Institute
www.mpi.org.au

Australasian Institute of Mining and Metallurgy
www.ausimm.com.au

Mining companies in Australia:

Equinox Minerals Limited
www.equinoxminerals.com

Newmont Mining Corp.
www.newmont.com

Silver Standard Resources Inc.
www.silverstandard.com

Expedition leader

Can't decide which extreme sport attracts you most? Keen to work as a rock climber, cross-country skier, mountaineer or safari leader? How about heli-skier, kayaker, Antarctic sled trekker or windsurfer?

If your idea of a fun day out involves climbing mountains, tackling sub-zero temperatures, kayaking through treacherous waters then finally sleeping under the stars, you could be the perfect candidate for a career in expedition leading.

It is essential to have a consuming passion for – and expertise in – wildlife, wild vegetation and wild places.

They lead a nomadic life, to be sure, but expedition leaders get paid to look after other people on trips to some of the world's most exotic places.

Expedition leaders are responsible for ensuring that their group travels safely along the planned route, introducing them to adventure experiences, teaching them about the wildlife in the area and – this bit is pretty important – bringing them home in one piece.

They might also supervise activities like bicycle touring, bushwalking, canoeing and kayaking, caving, cross-country skiing, horse trekking, rafting, rock climbing or sailing.

Expedition leaders take people on trips to exotic outdoor locations internationally – like Patagonia, the Himalayas and even Antarctica. They also run adventure holidays to the more remote regions of Australia.

Most leaders are involved in planning a safe route, organising all of the equipment and food that will be needed and arranging travel and accommodation for the group.

Expedition leaders are responsible for the group's safety so must also plan for any possible medical emergencies and be able to administer first aid in remote conditions.

A safe group is also a harmonious group. It's important that the expedition leader can control any quarrels or potential personality conflicts and be able to communicate effectively. The expedition leader needs to try to keep up a high group morale throughout the trip.

Often they work with people who are cold, wet, hungry, tired and sore, and who are well outside their comfort zones.

Excellent people skills and strong leadership qualities are essential.

 athlete >>
expedition leader

Eric Philips – Expedition company owner and leader

Eric Philips is the owner of IceTrek Expeditions and runs guided expeditions to the Arctic and Antarctica. He has led five expeditions to the North and South poles, and across Greenland, Patagonia, Canada's Ellesmere Island and Iceland. In his travels he has visited every continent on earth and has also worked as a film presenter, motivational speaker and musician.

What do you do in your job?
I'm in my third year of guiding people on expeditions to the North Pole

and I've been professionally adventuring – that is, it has been my main source of income, for about 13 years.

Do you employ people to work for you?

Yes, from time to time, although I can't offer anyone a permanent job as a leader. At present, I have a trainee guide who will help me guide a North Pole expedition next month.

What is the role of a trainee guide?

It depends on the expedition. For this particular trip, you certainly need to have experience in surviving extremely cold conditions and there are some skills that are essential. For example, you would need to have considerable cross-country skiing and mountaineering experience. The trainee guide who is accompanying me is not a fully-fledged polar guide – you need to have done an expedition to the Pole yourself before you qualify, so this is a training expedition for him. Although he won't be paid for the trip, he will have most of his expenses paid. He is currently doing a wilderness first aid course, so this trip will give him the credibility and experience that he needs to continue in this line of work.

What's involved in a North Pole expedition? Are all the participants very experienced?

It's a cross-country ski expedition that takes about eight days and covers about 110 kilometres. We will have six people with us on the trip. We have actually taken people on this trip before who were

completely inexperienced in outdoor activities – you do, however, need to be pretty tenacious and motivated to complete the trip.

You also need to be physically fit and healthy. I would have concerns about anyone who was overweight, had injuries, or was experiencing emotional distress, for example. The trip is open to anyone who has the physical capacity and a strong desire to do it.

What is your background?

I trained as a teacher and did a Bachelor of Education, with a Graduate Diploma in Outdoor Education. I've always had a passion for the outdoors. I worked in a few schools and ended up directing outdoor education at Timbertop for five years. I left in 1992 and have been involved in expeditions ever since.

What's your main source of income?

I gave up teaching to become an adventurer. I make my living from organising and undertaking private expeditions. Sponsorships fund the trips and pay me a wage during the time I am preparing and undertaking the trips. I also earn a living from speaking, selling photographs and films of my trips and from guiding. I don't make a huge amount of money. I make about the same as I did as a professional teacher 13 years ago – that was around $45K.

Are there vacancies for professional adventurers?

Any professional adventurer – and there are only a handful in Australia – didn't get to that point by applying for a job. For me, it began as a passion to be involved in adventuring as a hobby for its own sake. It was its own reward. I've been lucky to work myself into a position where it is possible to support my family and myself.

Eric has led five expeditions to the Poles.

What advice would you give anyone interested in becoming a full-time adventurer or expedition leader?

Go out and get yourself as educated as you can because there are many obstacles. A lot can stop you from making it and you will need to have something that you can fall back on that allows you to earn a living while you are building up your experience.

If you are interested in the area, there are formal qualifications in outdoor education that would be very useful. You should also ensure that you have as many certificates and qualifications as you can. For example, become a qualified cross-country ski instructor, and a rock climbing instructor.

What personal skills are important in this role?

You will develop many of the personal skills you need as you go through the process of learning various outdoor skills. But one thing you do need is the ability to sell yourself – and your ideas – to attract business sponsors or potential expedition participants, because that's what funds the expedition. If you don't have that knack for marketing or PR or whatever you might call it, you will have a difficult time making a go of it. Many expeditions to remote places – such as deserts, mountains, places like Patagonia or the icecaps – are very difficult to get underway and rely heavily on sponsorship.

Is there an element of luck involved?

Perhaps, but luck doesn't come to you unless you go looking for it.

Is it worth it?

Definitely. I love what I do, I could not imagine doing anything else.

Careers in expedition leading

For those who are passionate about the great outdoors, it sounds like the ultimate job – but leading an expedition to an exotic location is very different to backpacking around Europe or taking a gap year.

The role of an expedition leader is a very responsible one. The lives of the people who have joined the trip depend on your expertise, your knowledge, your personal skills and your behaviour.

It's a tough gig, with many limitations and not much privacy.

So why do people do it? Usually because they're passionate about the wild and have a strong desire to spend their time in exotic and inhospitable places.

If you feel that way about the wilderness and, more importantly,

find out more

Check out Eric's company website at www.icetrek.com

in brief

Expedition leader

$$$	0–50K
age	25+
quals	Outdoor qualifications and certificates
hrs/wk	35–100
life–work	Often away from home for months at a time

you wish to share it with others, you may be suited to the life of an expedition leader.

Many expedition leaders are also passionate believers in the role of outdoor education in personal development.

An influential figure in this is a man named Kurt Hahn who was behind the Outward Bound movement. Hahn believed that the skills and confidence to deal with unfamiliar territory can be learned and that exposure to challenges – in a secure environment – fosters a belief in oneself and compassion towards others that will last a lifetime.

Do I have what it takes?

To make it as an expedition leader you need to have some special attributes.

Excellent health, a high level of physical fitness and expertise in numerous outdoor skills are crucial.

But the best leaders have other qualities that set them apart, such as:

- leadership skills
- teamwork skills
- psychological stability
- attention to detail – capable of planning for every step and every need expected on the trip, as well as making allowance for the unexpected
- belief in other people
- an easygoing personality – you must be able to work with a diverse group of people at close quarters for long periods of time.

If this sounds like you, get packing!

Who could I work for?

Most expedition companies are small affairs, with the owner the main expedition leader.

However, there are several larger organisations that offer opportunities for leaders – such as outdoor education specialists Outward Bound and adventure travel specialists World Expeditions.

Expedition leaders fall into the occupational category of recreation

officers – people who 'plan, organise and coordinate recreation facilities and programmes'.

In February 2005, there were about 7400 people working in this category in Australia, but that includes people who work in indoor tourist venues – such as galleries and wineries – as well as outdoor professionals.

The industry is 62 per cent male, the median age of workers is 41 and the average wage is $1020 per week.

Huw Kingston – Outdoor event promoter, expedition leader and adventurer

Huw Kingston owns a small business called Wild Horizons which runs large outdoor events in Australia such as mass participation endurance mountain bike events and kayaking events, as well as outdoor adventure holidays.

 business student >> steel salesman >> traveller >> outdoor clothing salesman >> expedition leader >> outdoor equipment marketing >> outdoor equipment consultant >> adventure business owner

What does your job involve?

It's very difficult to describe because it covers a number of areas. I organise adventure events and run tours and expeditions, bushwalking, skiing and mountaineering trips. I write and take photographs for a number of outdoor magazines and I also work with outdoor equipment in design and manufacturing – so my work is very multi-faceted.

How long have you done this sort of work?

I have worked in the outdoor industry for 20 years, in a variety of areas. When I started this business, there was quite a big focus on designing outdoor equipment, but that has dropped off a lot in the last four or five years. Now the events have become a much larger part of the business.

What sorts of events do you organise?

This year we organised the 10th running of the Nalgene Polaris Challenge. It was Australia's first mass participation mountain biking event, which is run in a different part of NSW each year. There are about 600 riders in teams of two who compete over two days to locate various checkpoints over a course that covers bush and mountain. They carry all their gear – tent and so on – and have to read a map to follow the course properly.

Is there money to be made from this sort of thing?

Like most businesses, it's a question of controlling your expenses and making sure that you price your product appropriately. I've survived quite well with this sort of work.

Is there anything else to consider from a business perspective?

First, the participants need to have an enjoyable experience, with lots of support from us.

Secondly, there needs to be a community benefit – so we will make sure that we do some fundraising for the local area and that we employ local people to do catering for the event rather than bring it all in from outside. Small towns can make $10 000 to $15 000 from hosting our events.

Thirdly, the event has to be sustainable, with minimum impact on the environment. We don't want to damage the place that we have come to experience and we want the community to be happy to welcome us back again.

Do you employ other people to help you?

I have one person who works for me four days a week as an event coordinator and my wife Wendy does all the bookwork. Like most outdoor industry companies, we employ people on contract to work on individual events. But I don't usually employ anyone to do the expeditions, ski tours and other trips we run. That is the work that I want to do myself. There is no point paying someone else to do what you love doing yourself.

What do you like best about what you do?

It has really boosted my passion for the outdoors. I love the fact that I have the freedom to decide the direction of the business and what I am going to do next. My job takes me to beautiful parts of Australia and the world and I am able to work the business to allow me to have two, three or four months off each year where I pursue my own expedition somewhere. For the last seven years, I have been running a personal expedition where I do one trip per year kayaking, skiing, walking and cycling between each state capital.

What do you like least?

Like any job, there is an element of paperwork and dealing with tedious and often unnecessary bureaucracy.

Does this career allow you to achieve the life–work balance that you would like?

Yes and no. In one way, my work is my life, it's the way I earn an income. But because I have based the business at home since I've been working for myself, it can be hard to relax at home. I have no boundaries between my life and my work. It can be a bit suffocating sometimes, because you are always working, or there is always an element of work in your time off. My wife has been very supportive, but it can be hard on her. It is always difficult when you are away from the person that you love for long periods.

What advice would you give to someone who wants to work in this industry?

Nothing I have done has been planned so it's not like anyone can follow my particular career path – but if you know what you want to do then it is just a matter of persistence. Don't give up until you find a way to make it happen.

Skills you might not have thought of can be useful. I'm not a great believer in formal education – I don't learn well that way myself – but I did a business degree when I left school. Now I know how to budget and plan, and that's been crucial.

As for the events, sure, I get to go out and spend a lot of time planning the course and mapping and riding through the bush trails – that's all part of it and I don't deny that it's great. But the reality is, whether you're selling computers or selling events, the majority of the work that you do is just grinding away dealing with bureaucracy and the business side of things.

find out **more**

Check out Huw's company website at
www.wildhorizons.com.au

Get started!

No reputable travel organisation would entrust their clients to an expedition leader who didn't have extensive outdoor experience. But while experience in running your own trips is important, it won't earn you a wage on its own.

Expedition leaders are employed for their ability to organise and plan for other people and at the same time ensure that these people have a rewarding experience.

Courses and study options

Leaders are expected to have current first aid qualifications, ideally in Wilderness First Aid, which is available at many TAFEs across Australia. This covers specific topics for outdoors professionals who may need to provide first aid in remote areas, an hour or more from professional health services. You'll learn how to cope with extended patient care time, extremes of environment and the need to improvise resources.

Private colleges as well as TAFEs offer varying levels of outdoor leadership and outdoor education training courses, from Certificates II, III and IV to Diploma level.

Though not 100% necessary for work in the outdoor industry, formal teacher training in physical education or sports science is well regarded. If you're a qualified sports teacher or similar you might also find it a valuable safety net to fall back on while you're trying to break into the adventure field. It will also show employers that you're a people person and have training in the right sort of skills.

Some degree courses that might give you a head start in outdoor adventures and expedition leading include a Bachelor of Business Management (Leisure Management), Bachelor of Education (Physical and Health Education) and Bachelor of Sport Tourism Management.

Other useful qualifications that can be achieved include instructor certificates or proficiency certificates for various sports – such as skiing and kayaking.

The exact qualifications you'll need will vary from company to company, so make sure you do some research before you apply. For example, expedition leaders at outdoor education specialists Outward Bound must have qualifications and/or experience in the following areas:

- first aid
- safe leadership
- swift water rescue
- rafting and canoeing skills in both flat and moving water
- rock climbing and abseiling
- vertical rescue
- bushwalking and navigation
- search and rescue caving
- minimum impact camping
- ropes course facilitation and construction
- four-wheel driving
- adventure-based learning
- group management
- conflict management.

The most important qualification, of course, is experience – which means being prepared to do your time as an assistant guide or leader for very little pay, if any, just to get runs on the board.

If you're not already involved in outdoor adventuring, it's time to start. Scouts and Guides are good training grounds for younger people. In senior years of school or after leaving school you should consider joining a local group involved in bushwalking, rock-climbing, caving, skiing, kayaking or other outdoor activities. Your local council will often keep a directory of these groups.

Good luck in the great outdoors!

Outdoor adventures on the web

Adventure travel, training and jobs
www.adventurepro.com.au

Association for Experiential Education
www.aminwebsite.com

Institute of Australian Tourist Guides
www.australiantouristguides.com

Australian Sport Climbing Federation
www.climbing.com.au

Australian Ecotourism Association
www.ecotourism.org.au

Oz Quest Australia
www.ozquest.org

Professional Association of Climbing Instructors
www.paci.com.au

Professional Tour Guide Association of Australia
www.ptgaa.org

Wilderness First Aid training courses
www.wmi.net.au/wmi/

Youth Challenge Australia
www.youthchallenge.com.au

Outward Bound
www.outwardbound.com.au

World Expeditions
www.worldexpeditions.com.au

Extreme sports photographer

Athletes push the edge of what is possible – going up cliff faces, down mountains, faster, farther and higher. We know because we've seen the pictures. But have you ever wondered what's going on behind the camera?

Wherever there are extreme sportspeople, there are dedicated photographers dogging their tracks, snapping those amazing moments. Extreme sports photographers put their cameras – and often their safety – on the line. Their jobs take them to places most sane people fear to tread.

Extreme photographers must be in top physical condition and will usually require a combination of skills such as being able to abseil, ski, climb mountains and even parachute solo – all while taking photographs.

FAST FACTS

Neil Leifer's 1965 photograph of a victorious Muhammed Ali standing over Sonny Liston, floored and unconscious, is rated as one of the best sports photographs of the 20th century.

Mark Watson – Freelance adventure sports photographer

Sydney-based photographer Mark Watson first picked up a camera at age seven. Now he is one of Australia's foremost adventure sports photojournalists. His work covers key adventure events around the world and his photographs have graced the covers and pages of many of Australia's specialty magazines including *Inside Sport*, *Australian Mountain Bike* and *Outdoor Australia*. His work sees him underwater, hanging from cliffs, swimming through gorges and much more.

Where did the love of extreme sports come from?

I'm from Bells Beach in Victoria, so I have always been interested in surfing and other outdoor sports. My family always went on outdoor camping holidays.

When I left university, I worked a winter season at Charlotte Pass in the Snowy Mountains. The first year I was a lift attendant and the next year, I came back as a lodge assistant.

But you also had an interest in photography?

Having studied chemistry, physics and biology in high school, I enrolled in a degree course in scientific photography at RMIT in Melbourne.

I really started getting into photography when I was out there on the snowfields. That is where I started to shoot the adventure sports that I was interested in. I was out snowboarding with some really hot snowboarders when I realised what great pictures I could get, so I started taking my camera out with me.

So that's when you decided to become an extreme sports photographer?

Yes. You're always hearing that there is no money in wildlife photography, or that you cannot make a living from niche sports. I knew it would be difficult, but I decided to pursue my dream anyway.

Between the snow seasons I worked with a filmmaker in Kakadu and I later met my wife, who is English, while working my second winter

Check out Mark's photography at www.mwphotography.com.au

season. Then we lived in England for a couple of years and I put my photographic career on hold while I managed an adventure sports store. I wasn't taking photographs but I was still working in the area that I was interested in.

Finally, we moved to Australia and I started the photography business. It takes a while, but you can make a career in it – not just in sports photography, but in advertising and that kind of thing.

What do you actually do?

I shoot adventure sports images – mountain biker, snowboarders, adventure racers and sponsored athletes. The majority of the work I do is commissioned by orgaisations who want to be associated with action sports. For example, I do a lot of work with Red Bull and shoot the events they put on. On the freelance side my images are sold to magazines and advertisers.

What is a typical day like for you?

There is no such thing as a typical day – every day is different. In a typical week, I might spend a day or two doing admin work in Sydney, editing and filing photos and sending images to clients. The rest of the week is out in the field taking photos of athletes doing their thing, meeting clients and discussing the projects I am working on or even taking corporate portraits for annual reports and so on.

I spend about a third of the year away and I could be jumping out of a helicopter to take photographs of skiers in New Zealand one week and hanging from a cliff in the Kimberley trying to shoot cliff divers the next.

How much does luck play a part in getting the work?

You have to make your own luck. Photography works like the rest of the business world – it's as much about who you know as it is about providing a quality product.

So the business side of things is important?

Very important. My wife Sarah works in public relations. She taught me to go out and market myself. My successes have come from knowing people, meeting people and getting involved in the kind of activities I'm interested in. I don't have a Yellow Pages ad, but I will drop in from time to time and visit the editor of a magazine that I would like to sell pictures to, introduce myself and ask them what they are looking for.

There are a lot of very good photographers out there who don't make it because they don't run the business side of the job very well. It is essential to use every opportunity available to raise your profile.

What about clients?

As with any business you have to push a hell of a lot to acquire new clients. Building a solid reputation in the industry and being able to demonstrate that you can meet and exceed clients' photographic needs means you are halfway there.

A good example is Red Bull – they are marketing savvy and really understand the value of having a good photographer on board, and how much coverage they can get by working with a photographer who is willing to go above and beyond. When you come back from a job with elite athletes around the world and you hand them some blow-your-mind photos, they really know how to use them to get media exposure.

How much can you earn?

Most adventure sports photographers who are earning an income – and it is quite often a part-time income – will run their own business. You would usually start earning no more than about $10 000 in your first year. But when you are well established, your average turnover might be between $100 000 and $120 000 a year, of which a minimum $40 000 to $50 000 would be profit.

What's the most exciting event you've been involved in?

Probably adventure racing. You have teams of people racing across really remote and difficult terrain. It can involve anything from mountain bikes, white-water canoeing, hiking, abseiling, mountain climbing – you name it. The races can last for days and cover hundreds of kilometres.

I was hired to be the Australian adventure racing correspondent for *Triathlete* magazine, covering the Eco-challenge, which is a televised international competition.

I also covered such events as the International Raid X-Adventure series and the Land Rover G4 Challenge global adventure race. It

is sometimes crazy when you are working non-stop for days on end on only a few hours' sleep a night and camping in some of the toughest terrain imaginable. Just getting ahead of the competitors to photograph them can be a real challenge that can involve hiking, mountain biking, four-wheel driving or even hitching a ride in a helicopter to get to a vantage point.

What's the most challenging part?

You can't really be an expert in any particular area – you're jack of all extreme sports and master of none. Plus you're trying to keep up with the best people in the field, lugging around 20kg or so of incredibly expensive equipment.

Although the people that you are photographing will make allowances for you and you don't have to be an expert, you have to be able to hold your own. If you are shooting heli-skiing in New Zealand – you just can't risk going into that sort of territory without knowing what you're doing.

As an extreme sports photographer you have to be able to do everything that athletes can do.

There are not too many professionally trained photographers out there who can say, yes, I can get out of a helicopter, and I can snowboard down the same slope as these guys to get the photos. So that suddenly brings the field of competition down to only a few other photographers. In many extreme sports, a lot of the people doing the photography or cinematography are the actual extreme sportspeople themselves who have picked up a camera.

Read about Huw Kingston's
job in his interview on p121

in brief

Extreme sports photographer

$$$	10–90K
age	31
quals	Bachelor of Applied Science (Photography)
hrs/wk	20–100
life–work	frequently away from home

In a way that does give me a huge advantage. Because I have a background in scientific photography, I am used to looking at different aspects of photography and being creative about solving photographic problems. I can use technical solutions such as mounting a camera on a mountain bike and using a remote to try and shoot from the rider's perspective.

What sort of person makes a good extreme sports photographer?

You have to be very passionate, keen to try anything and occasionally a little bit crazy.

How difficult is it to balance your career with family and friends?

The main problem is the extensive travel, but then you also get to spend long periods at home to make up for it.

Did you get any advice from anyone about this particular career path?

Quite early on, I met adventurer Huw Kingston, who runs tours in the Himalayas and sea kayaking and mountain biking events. Huw showed me that you can make a living out of anything if you've got enough passion for it and if you put in enough hard work. He proved it by making a living from basically being an adventurer. He was an inspiration to me.

Any career tips for someone interested in extreme photography?

Don't give up – you need to be very persistent. You also need to develop solid photography skills and be passionate about adventure sports.

Careers in extreme sports photography

It's one of the most exciting and rewarding branches of photography – but it's very hard to make a living from it.

Hours can be very long, conditions can be uncomfortable at best, life-threatening at worst. So what's the appeal?

Well, extreme photographers have a passion for photography and a strong desire to follow extreme sports nuts and capture the outrageous things they get up to.

*I want to get into the action, to capture the essence of the
sport, to capture an image that makes you look a second time.
You know, you're reading a magazine, flipping through the
pages and suddenly you stop because the photo is so strong. It
transports you to that place and time so you can almost say that
you were there. That's the shot I'm after.*

Mark Watson, Adventure sports photographer, *Better Photography Magazine*, 2005

If photographers have the motto 'I like to watch,' then the motto of the
extreme photographer would have to be, 'I like to be there'. It's getting
out there to these amazing locations and being able to capture them
that appeals to these photographers.

Extreme photographers usually work alone, setting up photo shoots,
taking the shots, preparing them for publication and selling them to
magazines, newspapers and the sponsors of extreme sports events.

While it is important to be a technically competent photographer
– creativity and photographic skills are essential – it's also important to
be able to get to the spot where you can take that amazing photograph.

Professional extreme photographers would usually:

- run a small business – including the accounts, tax, administration,
  marketing and computer work
- network with people in the adventure sports industry
- obtain commissions from magazines or sponsors to photograph
  extreme sports events, negotiating expenses and fees in advance
- arrange the time and place for a photo shoot with the right people
  (event organisers or even the athletes themselves)
- travel to the location with all their camera gear and the appropriate
  equipment (abseiling ropes, skis, safety gear)
- spend hours – sometimes days – setting up and taking photographs
- process photographs and send a selection of the best images to
  the client
- look after administrative matters such as licensing, and understand
  racing rules.

Do I have what it takes?

Although experience and skill as a photographer is crucial, the
photographers who make it in the tough world of extreme sports
photography have many other skills.

Here are some of the skills that top extreme sports photographers should have:

- confidence in their environment – whether it be land, water or air
- physical fitness – they need to be able to keep up with their subjects!
- the ability to lug expensive and heavy camera equipment up a mountain or down a rope
- the ability to shoot quickly and creatively under extreme conditions.

Who could I work for?

There are around 7000 professional photographers in Australia. Most photographers are self-employed and their clients might include individuals, advertising agencies, graphic design studios, architects and corporations. A few work as press photographers.

There are more than a dozen regular sport and outdoor specialty magazines in Australia, as well as websites and club newsletters, so although it is a very small marketplace, there is a regular demand for good quality editorial images.

Companies who sponsor sports events are getting more involved in the marketing potential of these events and are often looking for quality images, although it is unusual to find a photographer on staff.

Magazine editors – particularly those from specialist adventure magazines – can be great clients for extreme sports photographers. They are certainly the ones you need to impress early in your career if you want to establish a track record of publication.

Lucas Trihey – Editor, Adventure Publishing

Lucas Trihey is the former editor of *OutDoor Magazine* and is now editor and co-owner of Adventure Publishing, which produces the annual *Australian Adventure Gear Guide* with equipment reviews and information for Australian bushwalkers, mountain bikers, paddlers and adventure racers.

He also runs a mountain guiding business and in June 2006 spent 17 days trekking solo across the Simpson Desert via an unexplored route.

q&a

How did you become editor of Adventure Publishing?

My background was in mountaineering. You end up in some pretty spectacular places and if you're off climbing on a little glacier high up on top of Mt Kenya that is actually on the equator, it makes for great pictures.

I studied photography at school in year 11 and just loved it. It became a bit of a passion. Once I started going on trips it naturally followed that I'd take photos and try to sell them. I am self-taught but the learning curve has been pretty steep. You spend so much time and effort and quite a bit of money getting to these dramatic places, only to have an editor knock your pictures back saying, 'Oh, not really sharp enough, sorry.' You soon learn to buy better lenses and spend a bit of time making sure everything is right.

It's just so hard to get to those places – you don't really get a second chance.

What advice can you give someone who wants to be an adventure photographer?

I'm an adventure photojournalist myself, as well as editing the Adventure publications. You can't be a photographer or a writer in this game unless you've got a background in adventure sports because you need the skills and the knowledge of the adventure side of things to get in and out of places. You can't go and photograph canyons unless you actually know how to abseil and how to look after yourself in a cold and wet environment and likewise you can't photograph mountaineering unless you're actually a mountaineer.

So you need a mixture of skills and they have to be appropriate.

How much competition is there for work?

There are a lot of amateur and semi-professional people who are really keen to get into this line of work. The industry is very prone to people undercutting other people because they just love what they're doing. Anyone going on an expedition somewhere will take a camera along and get a few good shots and they're just happy to have them published.

As an editor, it can be kind of hard to keep it professional because you've always got your amateur type people snapping a few pics and if it's a really exciting environment and if they've got any clues at all about the photography side of it they'll get a few really good shots – and if they're able to sell them, well, there goes that professional market. So it's a cutthroat industry in that way.

Any tips on how to learn about photojournalism for someone who wants to get into that area?

There are people who just take photos but there's not very many of them. If you can write as well as shoot images, it is a much easier way to make a living. You can approach any number of specialist magazines that are as interested in the words as they are in the photos.

When you start writing, just remember it's about telling a story. If you can focus on telling a good story and get some of the passion of it, it's going to work.

You're an editor who buys photographs – would you be more likely to buy a photo if it's sent to you with a story?

Yes, nearly always. In our particular magazine we don't have a gallery section where we just run photos. But quite often I get someone who can write really well but their photos are ordinary or they didn't have a camera with them, so I do have a list of photographers who I will approach to get an image.

Get started!

Still keen to try your luck as an extreme sports photographer? There are plenty of professional photography courses around which will give you an industry-recognised qualification.

There are also many small, private photography schools which offer short courses that can get you up to speed in things like exposure, shooting in low light situations and studio work (not something you usually need in extreme sports!).

If you plan on getting some part time work while you are studying – to save up for that expensive camera gear – aim for a related field such as a camera store, surf shop or adventure sports store, where you are more likely to meet people who have an interest in your field.

Most photographers emphasise the need to work as a photographer's assistant for a period of time to thoroughly learn the ropes before breaking out on your own.

Most sports photographers start out using very big, long lenses – 400mm and above – which allows for increased zoom. But some of the greatest photographers are now using fisheye lenses – 18mm and below. That means that they have to be very close to the action to get good shots, but the technique gives their shots a distinctive look.

Courses and study options

There's no shortage of photography courses on offer throughout
Australia. Depending on what prerequisites you need and how
much you're willing to pay, you could enrol in TAFE, university or
a private college.

Here are some TAFE courses that might fit your bill, whether
you're looking at certificate or diploma level. You'll need to check
the TAFE website for your state to see if any are on offer near you.

- Foundations of photography

- Photoshop basics

- Commercial photography

- Digital darkroom introduction

- Advanced techniques with digital cameras

If you're after a university degree, you shouldn't have trouble
finding one – but keep in mind that some universities need to
see your portfolio before they can admit you. Here are some
examples of degree courses in photography.

- Bachelor of Arts (Photography), RMIT University

- Bachelor of Arts (Multimedia), Charles Sturt University

- Bachelor of Photography, Griffith University

Then there are private colleges across Australia that run all kinds
of art and design courses, from short workshops of a day or two
to longer courses that result in professional accreditation.

These are just a few of the private colleges you can study
photography at.

- KvB Institute of Technology, Sydney

- Australian Centre for Photography, Sydney

- International College of Professional Photography &
  Multimedia, Melbourne

- Australian Academy of Design, Melbourne

- Queensland College of Art, Brisbane

Working as a freelancer – as most photographers in Australia do – can be very demanding and sometimes lonely, and you need to have the skills and knowledge to arrange and run a professional shoot and manage your own business. These things are best learned on the job as an assistant to someone already working in the industry.

If you're not already engaged in an outdoor lifestyle of some kind, now's the time. Build up your skills in snow, water, mountaineering or climbing sports.

You should also have a keen interest in photography from an early age. Most (but not all) professional photographers start shooting as keen amateurs before they leave school.

Join an amateur photography club – there is one in most local and regional areas throughout Australia. It wouldn't hurt to arrange a subscription to the extreme magazine that interests you most – check out newsagencies to work out which one best suits your style.

You can also check out the websites of photographer's associations, extreme sports publishers and groups and the personal sites of some of the world's great extreme photographers.

Photography on the web

Australian Institute of Professional Photography
www.aipp.com.au/home

Australian Commercial and Media Photographers
www.aipp.com.au/home

Media, Entertainment and Arts Alliance
www.alliance.org.au

Amateur and Professional Camera Clubs Listing
www.buy-n-shoot.com/cclubs.asp

Extreme Group
www.extrememediagroup.com

Extreme Sports
www.w1ld.com

Adventure Sports Online
www.adventuresportsonline.com

International aid worker

You've got to be a certain type of person to want to stick around in a war zone.

Roles working directly in war zones usually need people with very practical skills, such as trained health professionals, engineers, electricians, builders and people with experience in logistics who are able to coordinate food, aid and basic water and sanitation supplies.

FAST FACTS

The amount donated to charities by Australians has increased by about 60 per cent in the last seven years. In 2004, about 90 per cent of adults donated an average of $424 to charity.

Then there are the more administrative and support-related jobs such as translators and negotiators who pave the way for the delivery of supplies, and media coordinators who try to get the story out to ensure that donations reach the right area.

James Elder – Aid worker, UNICEF

James is 35 and has been working for The United Nations Children's Fund (UNICEF) in Africa for four years. For the past two years he has worked in Zimbabwe as head of communications and information media. The previous two years were in Angola (with a short emergency stint in Darfur).

He says his main ambition at the moment is 'to not get kicked out' of corruption-filled and strife-torn Zimbabwe.

What does your job involve?

I'm a press secretary for UNICEF in Zimbabwe. I field questions from media, stoke interest in various issues from AIDS to aid, put out fires (this is particularly relevant in a country like Zimbabwe, where the government views information as particularly sensitive and where mistakes can be very costly – closed access to your programs and the like), write speeches, help formulate policy, write features for foreign press, fundraise and so on.

What were you doing before this job?

I spent 11 years in the media, first as a writer on a Sydney magazine, then editor, then three years as a freelance journalist writing for everyone from *The Australian* to *Ralph* magazine, *Cleo* to *Good Weekend*.

What do you like best about what you do?

It's a cliché, but it's accurate – I do what I do because through this work I can make a difference.

There are days when you jump for joy at a breakthrough that can be translated into real impact for people at the bottom of this world – mostly children in faraway places. But then there are times when you shed tears for what failure really means to so many people who are beyond statistics but who actually touch your life from day to day. This job, the impact you can have, is an absolute privilege.

But at the end of it all, I live in amazing places where each day is an adventure – both physically and psychologically. This, of course, can be exhausting, but I wouldn't have it any other way.

What do you like least?

Power cuts, water that puts you on the toilet for days, the insight into what could be done for very little money, horrendous hardship on every street corner, the deathly reality and trying to escape it … and not being able hang out with friends at the beach.

What's been the focus of your job recently?

Selling to world media the idea that Zimbabweans deserve more than the world's sympathy – they deserve the world's support. I want to broaden the debate from politics to the people *affected* by those politics. One in three Zimbabwean kids are orphans, one in five adults are HIV-positive. So they are suffering on multiple fronts.

The tricky part was making the sell without offending the government, or appearing naive to the reality of politics in Zimbabwe. But we pulled it off, the debate has begun, and donors have dug deep.

What is the life–work balance like in this job? Is it hard being away from friends and family?

My wife works for the International Organisation of Migration (IOM), so she and our two children are here with me, and actually the life–work balance is brilliant. Indeed, it's one of the huge advantages of this life that we never envisaged. Though we have very limited social lives compared to Sydney, we have lots of time for our two children, Jasmine and Sebastian.

When my wife and I aren't working, we are with our kids, exploring Africa, playing and hanging out. Also, you have people working for you here – so any non-work time is free time. (No cleaning!)

Having said this, it hasn't always been this easy. In Zimbabwe you can still have a decent standard of living, but in Angola there was endemic malaria, depressing poverty, no entertainment, no space, lots of landmines and lots of bitterness …

There is no way I could have done this without my wife, Nicola. She not only offered critical support in a place like that, but she made the most of what it was. She found thrills in some of the madness, strength in that which would have worn most others down.

It takes a very special woman to live in a place like that – particularly one who is pregnant and with a toddler! – but she took all the positives out of it that she could, and dealt with the negatives in an amazing way.

What personal attributes are important in this role?

There is such an array of people in the UN, so it's impossible to say. Certainly I have a fairly large personality, which can be a shock to many cultures, but really anyone would fit in so long as they are open minded and willing to deal with conditions that most 20-year-old backpackers would reject!

Wanting to do something about injustice doesn't hurt either.

Does the work live up to your expectations?

Yes – and goes beyond them. I adore the work. At the risk of sounding like a wanker, I feel as though doors of perception have been opened for me to a few of the issues that are really going to matter on this planet over the next decade or two.

It's good to be in a place where there is no such thing as a simple answer, where more knowledge and understanding leads you to some conclusions, but ultimately introduces yet more pertinent questions.

I have a passion for it, for the children, for the whole idea of working for the UN. It ain't perfect, but it's more significant than anything else I could imagine doing.

UNICEF supplies shelter, sanitation and food in Darfur, western Sudan.

As well as all this, my wife and I have savoured so much of our children's early years. We aren't running around madly on the weekend seeing people – we are together, in a park or a pool, painting or climbing … It's been a joy and a massive bonus that I never expected.

What is something you had to learn the hard way?

I started my career with the UN in Angola – a country coming out of 30 years of civil war, where social services were devastated, death was on every street corner, housing was awful and as expensive as Tokyo, there were millions of landmines, and I didn't speak the language. On top of that, I was going from being a freelance journalist on Bondi Beach to the world's biggest bureaucracy – in an office ridden with malaria. I learnt everything the hard way.

What sort of work can people be employed to do?

A day at the office can mean many things to UNICEF staff. It could mean talking with a 14-year-old former child soldier about their experiences, or finding funding for vital supplies for children during an emergency, or dedicating each day to efforts to eradicate a killer disease.

How can someone apply for an overseas aid position with UNICEF?

Our Australian office is devoted to fundraising. Field positions are recruited from New York.

What sort of experience is important for someone in your position?

Experience as a journalist is critical. You need to know a good story, and how to tell it. Then try and jump the fence to the UN.

I spent three years backpacking, making money from bar work, restaurant work and walking dogs. It kept me on the road and gave me a deep love for travel, foreign landscapes, ideas and people ... So working for the UN is a bit like backpacking, in a way. Only, once you find yourself in an awful place, you can't just move on.

Any tips for preparing an outstanding job application?

Job experience, then more experience.

You also need to have a real understanding of who you are applying to. Have some ideas ready to give them about what you'd do if given the job. Don't just tell them how brilliant you'd be – show them.

Do you have any tips for handling an interview? What do you think got you the job?

Because I was applying for the UN Children's Fund, I started by saying how much this job offer made sense because I had newly become a father. The interviewer – a brilliantly cerebral (and, I would later find, childless) man – said, 'Yeah, yeah, blah, blah, blah the children ... Talk to me about strategy.'

I very quickly realised that the UN is not looking merely for people who think the issues are important, but for people who could actually grasp the issues, act on them, and generate results for children. I quickly started talking about how I'd sell children's issues in Angola to world press and donors. Concrete things.

What is the single most important thing you think you can do in this role?

Generate funds and publicity for the world's most disadvantaged children. I'm not desperate to climb any career ladder, but I remain naive enough to think that if I do my job well, then things will take care of themselves.

Find out about other aid worker jobs in Career FAQs *Save the World.*
www.careerfaqs.com.au

Media officer, United Nations

$$$	from local salary – $1K plus expenses – $80K+
age	35
quals	community work experience
hrs/wk	55
life–work	away from home and in places where communication can be poor for long periods of time.

Careers in international aid

Aid work is incredibly valuable and life-affirming, but there are plenty of people in need all over the world – so why would you choose to work in a country where you are in constant danger?

The answer is that most of the work available in charity organisations is not on the front line, but in coordinating resources and driving fundraising.

But the few people who do work at the coalface of aid work are passionate about the work that they do. Like most people who work in the community sector, they prioritise the importance of worthwhile work that will give their own life meaning.

In 2004, Australian individuals and businesses donated $11.5 billion to charity in goods and services, money and time, with $5.7 billion of this coming from individuals.

The lack of resources can be extremely frustrating. Money is tight, and budgets are tiny. Wages are usually low, the hours are long and the circumstances very trying.

But while they may not be well paid, the big perks of this job are a squeaky-clean conscience, and that warm fuzzy feeling you get when you've done a good deed – every day.

Do I have what it takes?

To really help people in a conflict-ridden environment, you need some pretty special characteristics:

- a compassionate nature
- the ability to not be intimidated
- the ability to see the big picture
- strength of mind
- negotiation skills
- resourcefulness.

A top-level aid worker is an ace bureaucrat, a logistics superhero, and a person who cannot be intimidated.

Who could I work for?

Aid is big business around the world. If you're interested in this field, you should have a good understanding of which major agencies operate in the region where you would like to work.

Most major aid agencies advertise vacancies on their websites and usually also include recruitment guidelines and tips.

Australian aid agencies generally have more vacancies for volunteers than they have for paid staff – but working as a volunteer is by far the best way to get into the area.

Ali Mackay – Volunteer coordinator, Oxfam Community Aid Abroad

Ali is the office manager and volunteer coordinator for Oxfam's Sydney office. We spoke to her about what sort of employment is available in Australian aid agencies.

Can you give me a little bit of background about your own career?

In my first real job I coordinated the first ever course that trained adult Indigenous Australians to work as professionals in early childhood. Since then I have worked in various community organisations – I basically followed my heart. I have worked for the **YWCA** doing community development, for Australian Volunteers International in a local **NGO** in India, with the education department, and now for Oxfam coordinating the volunteers.

Are there many opportunities for people to earn an income in aid work?

Actually, I don't think a lot of people get that chance. On the one hand, that's a very good thing, because more and more these days the people in communities in need are driving the projects themselves. It is less likely for Westerners to get positions in the field now, because we don't go and do that old-fashioned 'let's fix them up' kind of stuff.

What advice would you give someone who wanted to get into frontline aid work?

Coming in as a volunteer certainly increases your chances of getting work in the sector, but it's no guarantee. You still have to have some serious experience on the ground, and to understand the roles that you're taking on.

glossary

YWCA means:
– Young Women's Christian Association.

NGO means:
– non-governmental organisation.

A lot of the roles within the sector are not necessarily about saving the world. For example, it might involve running an event very efficiently. You could do that in the corporate world, or you could do it in the not-for-profit sector – the skills are much the same. But the money that you might raise from an event like **Trailwalker** goes to support our project partners in the field.

So there are more opportunities in, say, fundraising than there are in frontline aid work?

The things that we do mostly are project management, fundraising and events coordination. Most aid work is about providing support mechanisms for people who are actually in the field doing the work. In aid work we make sure that opportunities exist for everybody on an ongoing basis. Directly working in the development sector is probably only one way to be engaged in saving the world.

What sort of qualifications would help if you wanted to work at grassroots level?

To work in the field, most of the roles require really practical skills – people who are trained as health professionals, as teachers, as electricians, as builders.

I see so many fabulous people come through here who are doing their Master's in international development and so on. That's valuable, of course, but what they really need in third world communities is ongoing food production, sustainable agriculture, for houses to be built, and kids to be educated – the practical stuff.

glossary

Trailwalker means:

– an Oxfam fundraising event in which teams of four have 48 hours to complete a 100 kilometre trail.

Is there a certain type of person that is more suited to working in aid?

People who come in with a well-developed sense of social justice are going to do a really good job. Part of that is the very pragmatic reality that the sector is really committed to making sure that as much of the money as possible goes into the field, so resources are tight ... Hopefully, though, if you've got a sense of social justice, you'll be able to put up with a slow computer.

What's the pay like?

It's generally lousy. It's not good, unless you get up into senior management level where it probably is a lot better, but it doesn't really measure up against other areas of employment for the responsibility that you have.

What about the working conditions?

The working conditions are relatively simple in terms of infrastructure, but you generally work with a really supportive team of people. That's what makes the difference.

Are there other benefits?

I work for an organisation that people respect and that's a really good feeling.

Get started!

Working on the front line in aid work does not follow a traditional career path. It's also quite difficult to get your foot in the door. Most aid agencies operate globally so they have international workforces. You compete with people from all over the world when applying for a job and there is intense competition for locally-based roles.

In 2003, over 13 million Australians (86 per cent of adults) belonged to at least one non-profit association. Just under one million held office in a non-profit organisation.

Ideas cannot feed people and being kind or having strong views about poverty is not nearly enough. Agencies look for skills and knowledge that are going to be useful on the front line.

Skills that are in high demand include those of health and medical professionals, logistics experts, food professionals, engineers, agronomists and people with language skills.

Sometimes administrative skills can be very useful – such as accountants or even IT specialists, but these are less common requirements on the front line.

Professional work experience, experience with grassroots charity and community work in your local area and formal qualifications in a high-demand field are the best ways to make sure that you are qualified for the job.

Australia's non-profit sector is of a similar size to that of the United States. During 2004, 6.3 million Australians (41 per cent of adults) volunteered a total of 750 million hours of labour to non-profit organisations.

Courses and study options

If you're keen to work in international aid the most useful TAFE courses and university degrees are practical ones in education, languages, building and construction, engineering, nursing and health care.

However, there are more and more courses springing up specialising in international development and humanitarian aid. You can do these either as an undergraduate or postgraduate student, which certainly won't hurt your chances, so long as you put time and effort into volunteering and getting work experience too.

Here are some examples.

- Bachelor of Social Sciences (Human Services), Queensland University of Technology

- Bachelor of Arts (International Studies), Deakin University

- Graduate Certificate in Peace & Conflict Studies, University of Sydney

- Master of Development Studies, University of Melbourne

You can't become a front line aid worker straight from school or university. Aid agencies all stress that you need a reasonable amount of maturity and life experience. Professional skills in health, building and teaching and other practical areas are in high demand, so occupational experience is an advantage.

Another important criteria is experience in community work. The community sector in Australia operates in quite a different way to the business world, and time spent either as a volunteer or paid worker in a community organisation will usually demonstrate that you are able to use diplomacy and negotiating skills in a resource-tight environment.

Try to travel to developing countries, perhaps in Asia, Africa or South America. Travelling on a budget to these countries will give you first-hand experience of life in a poorer nation. Make sure that you can cope with this – ideally, you should love it! This will help you decide whether you would enjoy working in developing or conflict-torn countries.

If you think you have what it takes to be a front-line aid worker, you need to ensure that you have practical skills as well as community experience. Professional qualifications in health, building or teaching will be far more useful and likely to ensure a job on the ground than studying international development or other theoretical studies.

You should start researching aid projects and understanding the world of international aid and development. One excellent resource is the magazine *New Internationalist*, available by subscription.

A good knowledge of world affairs is important and you should keep an eye on international news – for example, you could watch SBS World News or read the international news sections of *The Australian* or the *Australian Financial Review*.

find out
more

Check out the *New Internationalist* website at www.newint.org

Not-for-profit workers account for 6.8 per cent of Australians in employment, making a contribution almost twice as large as the entire economic contribution of Tasmania.

In 1999–2000, Australia's non-profit sector contributed $42 billion to the national economy. This was equivalent to the contribution of the mining industry.

You might even consider spending a year with an organisation such as Australian Volunteers International. This gives you the chance to work as a volunteer on a project in a developing country, where you are given training and support.

Australian aid on the web

Aid Workers Network
www.aidworkers.net

Australian Government Aid Program
www.ausaid.gov.au

Australian Volunteers International (AVI)
www.ozvol.org.au

CARE Australia
www.careaustralia.org.au

The Fred Hollows Foundation
www.hollows.org

International Women's Development Agency
www.iwda.org.au

Médecins Sans Frontières
www.msf.org

Save the Children Fund Australia
www.savethechildren.org.au

UNICEF Australia
www.unicef.org.au

World Vision Australia
www.worldvision.org.au

About 24 000 Australians move overseas to live and work each year.

Ready, set, go for it!

What qualifications do I need?

So you've decided which extreme career is for you. But wait! There's something else you may need to think about: getting the necessary qualifications.

For some jobs, that might just mean bagging a traineeship, getting work experience, or doing a some other related job with lower entry requirements. For other jobs, you may need to do a university degree, TAFE course or industry-based course.

The fact is, to get into any of these positions, you will need to convince your potential employer that you are completely dedicated to that career path. (After all, *you* wouldn't want *your* skydiving instructor to be 'not really into it', would you?)

The following table shows the qualifications of each of the interviewees in this book. You'll notice that some interviewees have many qualifications while others had all their training on the job.

Job	Formal qualifications
Racing car driver	Automotive Mechanic; International Circuit Racing C (ICC) licence
Speed skier	n/a
Big wave surfer	n/a
Skydiving instructor	Minimum 600 jumps, plus numerous other specialised qualifications
Shark handler	Scuba-diving certificate, boat licence
Crocodile farmer	Animal handling or agricultural science qualifications
Aerial musterer	Pilot's licence
Clearance diver	Navy course
Underground blaster	Various licences
Expedition leader	Outdoor qualifications and certificates
Extreme sports photographer	Bachelor of Applied Science (Photography)
Aid worker	Community work experience

Career FAQs qualifications survey, 2006

Generally, if you're pursuing a career in a highly skilled profession you'll need at least an undergraduate degree from a university, and maybe a postgraduate course. Trades and more 'hands-on' jobs often require completion of a certificate or diploma at a TAFE institution.

Then there are private colleges and specialised schools that provide courses for jobs across the board. These are sometimes referred to as industry-based courses because they involve more work experience. Depending on your area of interest, you might complete a certificate in first aid, photography, occupational health and safety or carpentry.

Search the Internet for courses in your specialty or check out the boxed sections 'Courses and study options' in the previous chapters.

How do I get that job?

Once you've decided which extreme career suits you – and have a good idea of what you're getting yourself into – you're ready to get on with it.

Start by learning as much as you can about the field before applying for jobs. The interviews in this book are a good place to start – they show you the perspectives of people in the jobs and their employers, and refer you to other useful websites and resources.

Getting your foot in the door is the next challenge. There might be specific recruitment processes for particular careers. However, there are other ways to get a job that you should keep in mind.

How are people recruited?

People get started in extreme jobs in different ways. Recruitment tips are useful for landing your dream job, a related entry-level job or a work experience/volunteer position in a particularly competitive job. The main ways to get into these positions are:

- applying for advertised jobs

- cold-calling

- networking

- showing off

- getting an agent

- doing traineeships and apprenticeships.

Apply for advertised jobs

One way to get the job you want is to keep up to date with the jobs advertised in newspapers and online.

Newspapers are the tried and trusted place to look for job vacancies. Though you can find jobs any day of the week, many Saturday papers have a specific careers section. The CareerOne section of *The Weekend Australian* advertises many vacancies Australia wide as does My Career in *The Sydney Morning Herald*. Local community newspapers also list job vacancies.

Internet job ads are growing in popularity with employers. Jobs are usually posted for a month, rather than a day as in newspapers.

You will be able to browse ads in newspapers and on the Internet according to the sectors they are divided into, such as health, IT or entertainment.

If you want to work for a government department or agency, look at job vacancies in the *Public Sector Gazette*. All government job vacancies can be found at the Gazette's website.

For specific jobs in obscure or very small sectors, be sure to look for job ads in industry association websites and newsletters. Your dream job as a lion tamer might be advertised in Big Top Monthly rather than a national or statewide newspaper.

www.seek.com.au

www.mycareer.com.au

www.careerone.com.au

Public Sector Gazette Online
www.psgazette.gov.au

Cold-calling

If sitting around waiting for the right advertisement isn't extreme enough for you, door knocking or cold-calling on potential employers might be more your style. This basically means that you visit businesses you'd like to work for and ask to speak with the manager (politely – leave the balaclava and camouflage gear at home). Then explain who you are, why you're there and what you believe you could contribute to the business if you become part of the team.

Most employers like initiative, and will appreciate your enthusiasm and commitment to working. And, let's face it, even if you don't get a job from it, at least you'll get to practise and perfect your interview skills.

This tactic is particularly good for any industry that doesn't have many job openings. If you've left a good impression, chances are you'll be the first person they think of when a job opening does pop up.

If you're planning on having a go at this, make sure you dress professionally and think of the person you speak with as an interviewer, valuable network contact and potential employer. Take copies of your résumé and references to each employer.

Networking

Sure, making friends is good – but making friends you can call upon later to exploit shamelessly is even better!

People you meet on your career path (at university, TAFE, college or in related jobs) might turn into valuable contacts and help you get your job somewhere down the line.

Networking is all about getting to know the big fish in the industry – and this doesn't only apply to aspiring shark handlers. If you're interested in being a racing car driver, throw on your networking helmet and head down to the track. If you're keen on being a big wave surfer, put on your networking Speedos and check out a surfing competition now and again.

Showing off

Whatever the field you dream of working in, there are always opportunities to strut your stuff. Getting your name out there will make you more valuable to employers in a competitive area, and give you some certificates to add to your portfolio.

Particularly if you're interested in the sporting arena, you should keep an eye out for competitions, scholarships and traineeships in your field. Never heard of an aerial mustering competition? Me neither, but you'd be surprised what you can find searching the Internet.

Getting an agent

You can benefit from having an agent if you work in entertainment, the arts and sport. Agents are the employment agencies of the entertainment and sporting world. They represent people with particular skills and can play a significant role in launching and guiding their careers, assisting them to get their names out there and keeping them informed of the job opportunities available.

Traineeships and apprenticeships

Traineeships or apprenticeships are a great way to start working in a job. You learn while working and observing experienced workers. And – here's a secret – you get paid for it. Afterwards you may be offered a job or you might apply for jobs at other places. Traineeships can be part of TAFE or industry-based courses.

Stand out from the crowd

Applying for jobs that are a little out of the ordinary can be a very competitive process. You will need to be prepared for the expectations of recruiters and present yourself in an attractive way. Now, that doesn't mean lying about your skills and accomplishments …
'I've raised *plenty* of crocodiles!' 'I have *heaps* of experience with explosives!' 'Of *course* I can eat my own head!' Not a good idea. You do, however, need to present yourself in the best possible light.

Create an outstanding application

If you want to get a great gig you need to know how to create a great job application. One of the first things you should do is take the time to plan how to best sell your skills, experience and ambition to potential employers.

The résumé

Your résumé is the summary of your career, education and achievements – academic, sporting and artistic – that relate to your aspirations. It must be presented in a clear and concise way for potential employers to be able to get an idea of who you are just from glancing at it – so try and keep it to a maximum of two pages!

You should have a résumé ready at all times and it's never too early to start putting one together. You'll make life easier for yourself by writing one now and then updating it regularly. There are many ways to put a résumé together, but all have certain things in common. You can see the main features of a résumé in the example below.

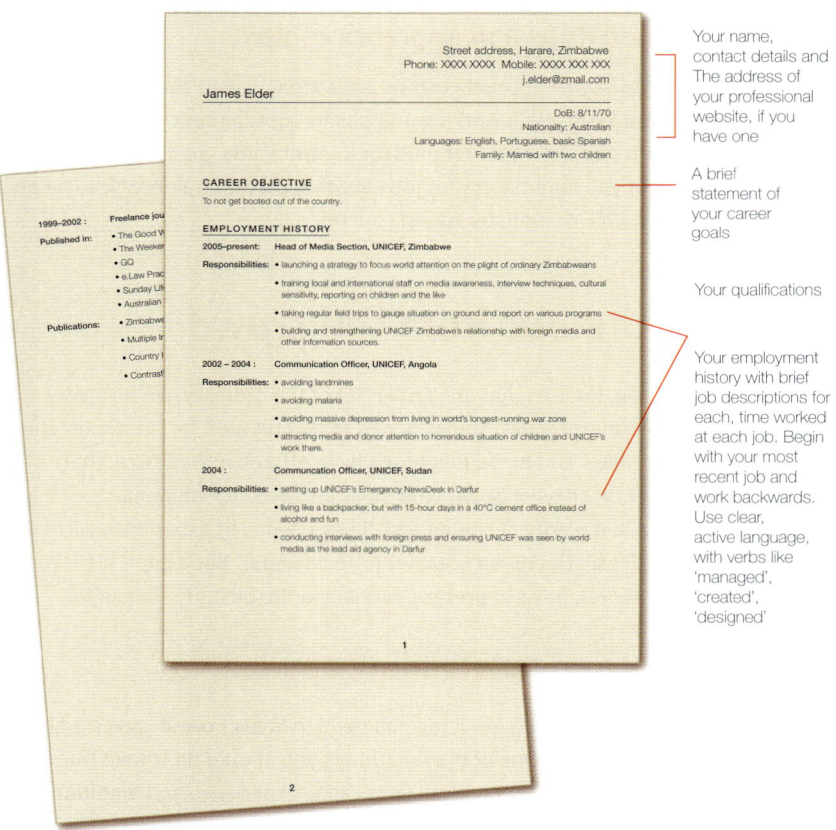

You can see a full-sized version of this résumé in Appendix 1.

The cover letter/email

Never send out your résumé to a potential employer without some sort of introduction. This means writing a cover letter and sending it along with your résumé. Keep in mind the fact that more and more recruitment is being done via the Internet, which means your cover letter may actually take the form of an email.

You should begin by expressing your interest in the advertised position, then summarising why you should be considered for the position. Be sure to address the employer's specific criteria, if there is any. And, last but certainly not least, don't forget to attach your résumé!

Whatever you do, don't send out a generic letter or email. Make sure every letter is tailored to the job you're applying for.

Here is an example of a well-written cover email.

Check out Appendix 2 for a cover letter version of the above cover email.

A cover email should be brief and to the point. If you'd like to give more description about yourself and your work history it is better to write it as a letter in Microsoft Word, and send it as an attachment in a short, polite email.

A cover letter should be one page long, two at the very most, it must be well formatted (white space is your friend), and you need to include your name and contact details as well as the employer's at the top of the page.

The portfolio

Your portfolio is a collection of some of your best work. Depending on your area of interest it could contain shots of your sporting achievements or copies of your own photographs, designs or websites. Take your portfolio to job interviews and wherever you'll be around potential employers. Having a portfolio allows you to:

- have something to take to interviews
- showcase your skills, abilities and achievements
- demonstrate the results of your work
- get into the habit of documenting your accomplishments
- create a personal database
- clearly present your career progress to date.

You can buy portfolio books from stationery shops. A good portfolio book can be quite expensive – often up to $100 – but you will be carrying it around a lot to show prospective employers, and there's nothing worse than having a battered book, or one where the pages fall out too easily. Who wants to see all their beautiful designs come crashing to the floor? Portfolios are a great professional touch, and are the first tool you need to sell yourself as an independent contractor.

The website

Check out Appendix 3 to see a sample website.

There are certain jobs in which having a professional website of your own is absolutely vital – particularly if you are running your own business. This also goes for professional sportspeople – it's a way of collating your achievements in a form that allows sponsors to see what you're all about.

If you're good with computers you could have a go at setting up your own site – but it's probably going to look more professional if you hire an expert to do it for you. You may need to make a small investment – but it will definitely be worth it in the long run!

How can I stand out at the interview?

Whether you're being interviewed for a job, a traineeship, a work experience position or as the defendent in a criminal investigation, you need to tackle it the same way: by working on your preparation, presentation and performance. The three Ps.

Preparation
Research the company
- Read or skim the annual report and any other information.
- Find out about the structure of the organisation
- Get to know the key positions and the names of people in them

Interview 3 Ps

Presentation
Be personable and dynamic
- Focus on your strengths – you'll be less nervous
- Develop a rapport with the interviewer/s
- Dress appropriately and be well groomed
- Speak clearly
- Make eye contact
- Smile, relax, breathe!

Performance
Find out about interview techniques
- Make sure you answer the question – don't get side tracked!
- Be honest in your responses
- Talking about mistakes isn't a bad thing if you show that you have learnt from them

Tips: The interview

- Be prepared. Research the industry and company thoroughly.

- Be on time.

- Turn your mobile phone off.

- Be yourself. Be truthful and direct, but try to put a positive spin to your answers.

- Make sure you can substantiate your information.

- Don't tell the interviewer about any bad experiences you may have had with other people or companies.

- Ask questions. This shows you're interested and that you've done your homework.

- Be polite and professional, and listen carefully to the questions.

- Don't ask about money at the first interview.

Sponsorship

In some fields corporate sponsorship may help you get ahead in your career. Sponsorship is a mutually beneficial relationship in which a company will financially support a sportsperson or team in return for the right to associate their name with the sportsperson.

This means the person may have contractual obligations to wear the brand's clothing or use the company's equipment, appear in commercials, attend press gatherings, take part in sporting or social events or sign autographs for fans.

Sponsorship can be a great way to get your name and face out there. When you're starting out you may be thankful for the equipment, services or management you're being offered. Just remember that you'll be associated with the brand, so make sure you like their image and genuinely approve of what they do.

If you're starting out and are showing enough potential to attract sponsorship, get some legal advice about the deal before you sign anything.

How can I start my own business?

If you're extra wily and determined, establishing your own business might be an option – particularly if your career is so extreme or groundbreaking that there are no traditional employers! Many of the people interviewed in this book are running their own businesses – sports photographers, expedition leaders, crocodile farmers, this mean YOU!

Check out Appendix 4 for our guide to starting your own business.

❛ Contrary to common perceptions, most Australian businesses survive for a considerable time. ❜

Around two-thirds of businesses are still operating after five years and almost one half are still operating after 10 years. It's good to know that less than 0.5 per cent of businesses went out of business because of what is referred to as catastrophic failure – bankruptcy or liquidation. And the failure rate has been falling since the early 90s (*Business failure and change: an Australian perspective*, 2001).

Richard Branson has a simple, easy-to-remember philosophy about setting up and running a successful business: 'People, people, people'.

> Surround yourself with people better than you, learn the art of delegation, have a passion for the business with real objectives and just give it a try.
>
> Richard Branson at the inaugural Meet the CEO event, UNSW, 2003

Good advice from someone who knows.

While owning your own business in Australia does not require formal qualifications, if you're planning to own a successful business, it helps to know your trade and a few of the rules of good business.

The business plan

Your business plan is essentially your official business strategy – it is an official record of your business idea, product and objectives. It's also an important document to take to the bank if you need a business loan!

Many websites have ideas about business plans and how to write them. Others have free samples and downloadable files, so be sure to make the most of the Internet for research purposes before you start writing.

Check out our guide to writing your own business plan in Appendix 5.

Industry programs

There are many industry courses available to teach you the essentials of small business ownership and operation. You can learn through formal training, mentoring and industry seminars. Small business management topics usually include:

- accounting procedures
- pricing and merchandising
- staff management
- fulfilment of legal obligations and legal rights.

The Australian Institute of Management (AIM) is a good source of information and they have courses relating to business management.

find out more

www.dewr.gov.au

www.ausindustry.gov.au

www.nna.asn.au/NEIS.htm

www.centrelink.gov.au

www.ato.gov.au

www.aim.com.au

www.linemanagement
.com.au

glossary

Flexible delivery means:

– off-campus, external,
online or distance education
learning. Because of
Australia's huge size, many
universities and colleges
offer flexible delivery as a
study method.

find out more

www.australian-universities.
com

www.tafe.nsw.edu.au

www.tafe.wa.edu.au

www.tafe.sa.edu.au

www.tafe.vic.gov.au

www.tafe.qld.gov.au

www.centralian.nt.edu.au

www.cit.act.edu.au

The Brisbane-based LineManagement Institute of Training is another organisation providing business training courses. Their course in small business management is specifically designed for people starting or running a business.

These are just two examples of industry programs for businesspeople in Australia – but there are plenty more out there! Find some local organisations and see what they have to offer.

University and TAFE courses

There are dozens of TAFE colleges and universities throughout Australia offering courses that will benefit anyone considering starting their own business.

TAFE offers a range of courses from entry-level statement of attainment courses through to Certificate IV in Small Business Management. Courses are designed to train new and established small business owners in the essential skills needed for successful business ownership.

Courses in small business operations are also available Australia-wide through the Open Learning Institute of TAFE's **flexible delivery** program.

Most universities offer undergraduate degrees in business, where students can choose to major in HR, marketing, management and so on.

If you already have a Bachelor's degree or are thinking of entering as a mature-age student, you might consider applying for a Master of Business Administration (MBA). Within this course you cover the key areas of contemporary management practice and can select subjects from courses in marketing, international business, business economics, accounting, finance and public administration.

Additional support

Centrelink funds a program called the New Enterprise Incentive Scheme (NEIS) that helps people on Centrelink benefits start their own businesses. In this scheme, you receive professional mentoring, advice and training on the process of starting up a successful small business.

For more sources of support when starting your own business see Appendix 6.

Tips: Starting your own business

- Talk with people working in the industry and invest in some industry trade journals to find out what's hot and what's not.

- Look for current and future small business incentives and determine which ones, if any, your business is eligible for.

- Conduct a focus group to find out how marketable or popular your business may be.

What can I do right now?

Exciting adventures. Exotic locations. Exhilarating life. And other things not necessarily starting with 'ex'.

If all this is *exactly* what you're looking for, you might want to start thinking of ways to break it to your friends and family. Remember how your mum hit the roof when you started that snail farm in third grade? Your new business in anaconda farming probably won't go down too well either.

It's all about the art of diplomacy. Remind your folks gently that they won't be complaining about the genuine crocodile-skin boots they've got coming. Explain that there are options – you don't mind Formula One OR dirt bike racing. And let's not forget all those free circus tickets you can give them to your amazing one-person lion-taming/fire-breathing/sword-swallowing act …

And remember – if it doesn't go down well, you could always threaten to do an arts degree. That'll show them.

Buzz words

banking	tilting an aircraft in flight
camera flyer	someone who takes pictures of other people's skydives
CDT	Clearance Divers Acceptance Test
dangling the Dunlops	getting an aircraft ready to land
face	the forward-facing portion of the wave where wave riding usually occurs
flexible delivery	off-campus, external, online or distance education learning. Because of Australia's huge size, many universities and colleges offer flexible delivery as a study method.
HR licence	Heavy Rigid licence. Holders are eligible to drive a vehicle with three or more axles, for example a bus or a truck.
jackaroo	an apprentice at a colonial station
Jaws	a beach on the Hawaiian island of Maui, famous for having the biggest waves in the world
jetsprint race	a competition where a two-person crew race their boat against the clock through water channels on a pre-determined course
jillaroo	a female jackaroo (see 'jackaroo')
NGO	non-governmental organisation
open cut mining	a form of surface mining. Minerals are extracted from the earth through a shallow open pit. Sometimes called open-pit mining or open-cast mining.
PNG	Papua New Guinea
scuba	self-contained underwater breathing apparatus. Divers carry a scuba tank on their back and breathe in through a regulator.
shuttle run	a demanding series of 20-metre sprints
stope	avoid created in the mining process
Trailwalker	an Oxfam fundraising event in which teams of four have 48 hours to complete a 100km trail
twin engine endorsement	an aviation rating qualifying a person to fly a twin engine aircraft
underground mining	excavation that requires tunnelling under the earth. Also known as sub-surface mining.
YWCA	Young Women's Christian Association

Appendix 1

Sample résumé

Street address, Harare, Zimbabwe
Phone: XXXX XXXX Mobile: XXXX XXX XXX
j.elder@zmail.com

James Elder

DoB: 8/11/70
Nationailty: Australian
Languages: English, Portuguese, basic Spanish
Family: Married with two children

CAREER OBJECTIVE

To not get booted out of the country.

EMPLOYMENT HISTORY

2005–present: **Head of Media Section, UNICEF, Zimbabwe**

Responsibilities: • launching a strategy to focus world attention on the plight of ordinary Zimbabweans

• training local and international staff on media awareness, interview techniques, cultural sensitivity, reporting on children and the like

• taking regular field trips to gauge situation on ground and report on various programs

• building and strengthening UNICEF Zimbabwe's relationship with foreign media and other information sources.

2002 – 2004 :   **Communication Officer, UNICEF, Angola**

Responsibilities: • avoiding landmines

• avoiding malaria

• avoiding massive depression from living in world's longest-running war zone

• attracting media and donor attention to horrendous situation of children and UNICEF's work there.

2004 :   **Communcation Officer, UNICEF, Sudan**

Responsibilities: • setting up UNICEF's Emergency NewsDesk in Darfur

• living like a backpacker, but with 15-hour days in a 40°C cement office instead of alcohol and fun

• conducting interviews with foreign press and ensuring UNICEF was seen by world media as the lead aid agency in Darfur

1

1999–2002 :	**Freelance journalist**	
Published in:	• The Good Weekend	• The Australian
	• The Weekend Australian	• The Sydney Morning Herald
	• GQ	• Ralph
	• e.Law Practice	• Qantas
	• Sunday Life	• Inside Sport
	• Australian Style	• Cleo.

Publications:

• Zimbabwe's Orphans and Vulnerable Children Baseline Survey, 2005 (editor)

• Multiple Indicator Cluster Survey, UNICEF Angola, 2003 (editor)

• Country Information Kit, UNICEF Angola, 2003-04

• Contrasting media coverage (Western versus Arabic) in Middle East politics, 1995.

2

Appendix 2

Sample cover letter

Jimmy Applicant
1 Alpha Avenue
Betaville NSW XXXX
(02) XXXX XXXX / 0401 XXX XXX
email: jimmy.applicant@chainmail.com.au

Ms Mary Employer
Employers and co.
333 Theta Road
Kappaville NSW XXXX

Dear Ms Employer

I am writing in response to your advertisement for a communications officer in Nairobi.

For the past four years I have worked throughout Africa as a communications officer for a not-for-profit organisation and thus consider myself well suited to the role you have on offer.

In 2000 I completed a Bachelor of Arts in Communications (Journalism/ International Studies). Since graduating, I have gained experience as a freelance journalist, writing for *The Weekend Australian*, *The Australian*, and *The Sydney Morning Herald*, among others.

In addition to this, I have functioned as a marketer, fundraiser and researcher. Having lived in countries racked by civil strife I am accustomed to working under pressure and well versed in the nuances of cultural sensitivity. I am comfortable dealing with the media and am familiar with all facets of work as a communications officer.

Your charity has an excellent reputation worldwide and I would appreciate the chance to meet your team. My résumé and portfolio samples are attached. I look forward to hearing from you.

Best regards

Jimmy Applicant

Appendix 3

Sample website

Example of extreme sports photographer Mark Watson's professional website.

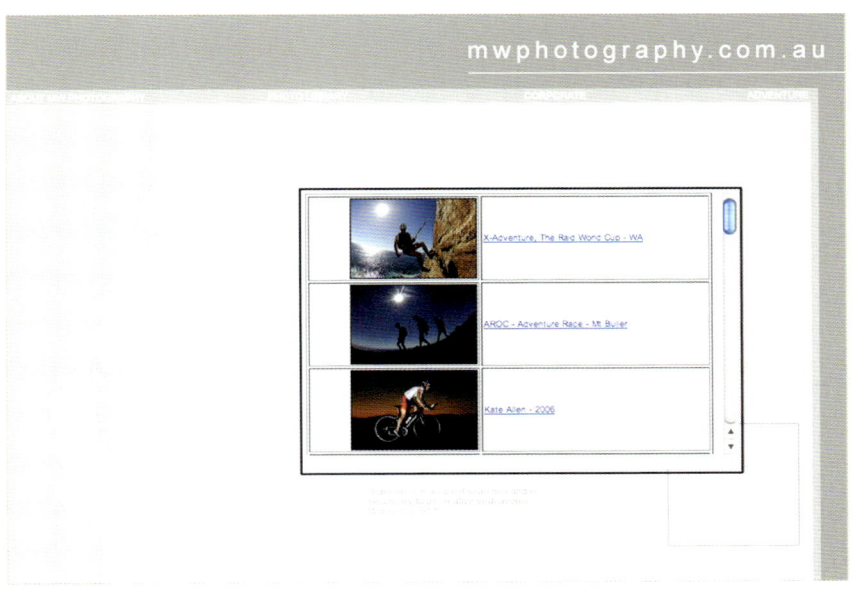

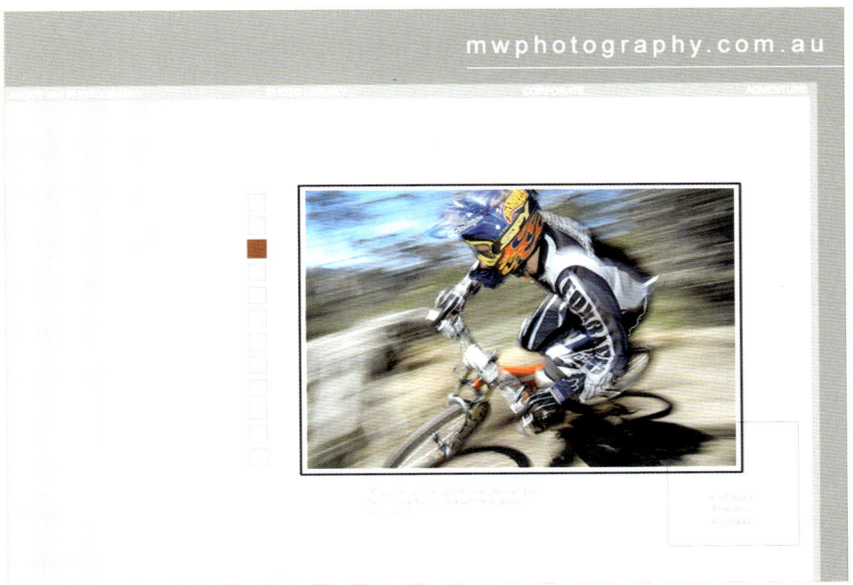

Appendix 4

Guide to starting your own business

Define your product
- What is it?
- What is different about it?
- What is desirable about it?

Decide what your target market is
- Does your target market consist of students, parents, retirees or young professionals?
- What products are available for your chosen market?
- Is it a specialty market? Do you have enough professional knowledge of the market?

Choose your equipment and office location
- Considering your product and audience, what would be best?
- Consider transport, fuel costs and maintenance
- Can you run the business from home at first?

Check the legalities
- What are the legal requirements, including occupational health and safety, environmental health and insurance?
- What are the legal fees and business establishment fees and you will have to pay?

Create a business plan
- Write down the nuts and bolts of your business
- Target audience
- Target market
- Competition (direct and indirect)
- Products offered
- Location
- Cost of renting/buying premises if necessary
- Machinery or technology needed
- Proposed marketing campaign/s

Check your finances

- How much will it cost?

- How much do you have?

- What bank can offer you the best deal?

- Devise best case and worst case scenarios and calculate what finances you'll need to cover the worst case scenario.

Revisit and refine your business plan

- Is there anything else that needs to be factored in?

Appendix 5

Guide to writing a business plan

Summary:
A brief summary of what your business idea is.

Product or service description:
An in-depth look at what you want to sell with your business.

Business concept:
What makes your business different from others?
Why will it be successful?

Market:
Who is it for?
Statistics – population, target demographics, short- and long-term sustainability and projected growth.

Competition:
What similar businesses are in the local area? How many? Why is your service different? Why is your service needed in this area?

Personal experience, knowledge and skills:
What skills, knowledge and experience do you bring to this business that will help it, and you, succeed?

Entry strategy:
How do you plan to promote your business for maximum effect? What is your marketing campaign for the first 6 months? How will you get people to (a) hear about your business and (b) visit your business as customers?

Operational and resource requirements:
What technology and resources does your business need to be viable? In other words – what will you need to pay for before you're properly set up? This may include a camera and camera equipment, skydiving licence, a short course in orienteering, a car or mobile phone, a pen to keep your crocodile in and so on.

What human resources do you need, if any? How much will staffing cost? How much will training cost?

Growth strategy:
How do you plan to grow your business? Do you plan to franchise? Do you plan to expand the concept to include other products?

Exit strategy:
If the business is not successful, how do you plan to minimise losses if you have to close the business?

Regulatory and legal requirements:
What legislation and regulations control the business environment? How do they apply to your business? How do they affect your business? Are they likely to change? If so, what effect will that have on your business?

Financial projection:
How much will you need to start this business?
How much will the business make?
Business figures based on merchandise costs, overheads, potential profit, number of sales/customers needed to attain that figure.

Action plan:
If this business plan gets the go-ahead, what's your next move? It could be buying a vehicle or trailer, purchasing equipment or implementing a marketing campaign.

Appendix 6

Sources of support for small businesses

Federal Government

AusIndustry

AusIndustry is the Australian Government's business program that provides a range of incentives to support business innovation. It is part of the Department of Industry, Tourism and Resources.

It has some 30 business products including innovation grants, tax and duty concessions, small business services, and support for industry competitiveness worth nearly $2 billion each year to about 10 000 small and large businesses.

To help customers with product and eligibility information, AusIndustry has customer service managers located in 26 offices across Australia, a national hotline and website, plus almost 60 small business field officers in regional areas.

AusTrade is another federal government agency that has been set up to help Australian businesses. The focus of AusTrade is the export market, so when your business has conquered the Australian market, AusTrade may be able to help break into some overseas markets.

Government departments

The Department of Employment and Workplace Relations provides access to online services and information, guiding you to employment information, government assistance, training, working conditions and Indigenous Employment Centres.

The Department of Workplace Relations and Small Business also operates Wageline (the telephone service) and Wagenet (the Internet service). Both provide details of federal and state awards for different industries and details of workplace agreements and certified agreements.

Australian Taxation Office

The Australian Taxation Office (ATO) can help with all the issues related to taxation – very important to get right from the start.

You can get information on Business Activity Statements, Australian Business Numbers (ABNs) and guides to record keeping – all essential for running a successful business.

State and territory governments

Each state and territory government has their own business advisory departments.

New South Wales

Business Education Network

www.nsw.gov.au/business.asp

Queensland

Department of State Development and Innovation

www.sdi.qld.gov.au

Tasmania

Department of Infrastructure, Energy and Resources

www.iris.tas.gov.au

BizTas

www.biztas.com/entrypoint/biztas

Victoria

Department of Innovation, Industry and Regional Development

www.business.vic.gov.au

Victorian Business Channel

www.business.channel.vic.gov.au

South Australia

Department of Trade and Economic Development

www.southaustralia.biz

Western Australia

Department of Industry and Resources

www.doir.wa.gov.au/businessandindustry/index.asp

ACT

Department of Economic Development

www.business.act.gov.au

Northern Territory

Territory Business Channel

www.tbc.nt.gov.au

State and territory government departments specialising in business advice

Each of these departments aims to facilitate their state's economic development by providing the means for business people to start and improve their business operations.

Notes

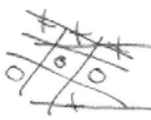

Photo credits

A big thanks goes out to everyone who generously supplied photos for this book.

Front cover: Frederic Carmel, stock.xchng

Chapter dividers

The big picture: Sasan Saidi, stock.xchng

Race car driver: Steven Johnson

Speed skier: Avnerr, Dreamstime.com

Big wave surfer: Taylor Hunt, stock.xchng

Skydiver: Andreas Jankowsky, stock.xchng

Shark handler: Tash Whitely, stock.xchng

Crocodile farmer: Carol Lam, stock.xchng

Aerial musterer: Jeremy Menking, stock.xchng

Clearance diver: Christian Schiedl, stock.xchng

Underground blaster: Paul Graham, stock.xchng

Expedition leader: Blazej Pieczynski, stock.xchng

Extreme sports photographer: Philip Jackson, stock.xchng

International aid worker: Kevin Rohr, stock.xchng

Ready, set, go for it!: James O, stock.xchng

Miscellaneous

Bryce Adams, Andrew Brigmond, Ross Clarke-Jones, Daron Cooke, Ian Dalton, Ashley Dowden, Tom Denham, James Elder, Jeff Hallam, Steven Johnson, Sandor Kapocsi, Pit Klad, Adrian Laing, John Lever, Luke McAdam, Michael Milton, Christine Nesbit, Christina Orr, Eric Phillips/IceTrek Expeditions, Tibor Szentmarjay, Mark Watson, Katie Weir, Isobel Wheeler

stock.xchng is accessible at www.sxc.hu

Books for every career you can imagine!

Available now at bookstores and on the Career FAQs website

Accounting
Accounting NSW/ACT
Advertising
Allied Health
Building & Construction
Design Professionals
Engineering
Entertainment
Extreme
Fashion
Financial Planning
Going Global
Hospitality
Human Resources
Information Technology
Investment Banking

Landscaping & Horticulture
Law
Law NSW/ACT
Law Victoria
Marketing
Medicine
Nursing
Nursing NSW/ACT
Nursing Victoria
Psychology
Public Relations
Save the World
Teaching NSW/ACT
Travel & Tourism
Weird & Wonderful

Coming soon

@gov.au
Accounting Victoria
Banking
Be Your Own Boss
Beauty & Fitness
Digital Media
Education
Industrial Design
Journalism

Publishing
Scientific Pursuits
Teaching Victoria
The Art World
The Sporting Arena
Working from Home
Working with Animals
Working with Children

Other 'expand your horizons' books

Going Global

Want to work overseas? The world's your oyster and Career FAQs *Going Global* your guide. You'll find interviews with Australians working in the UK, the US, Canada, the Netherlands, China, Cambodia, Sudan and more. We give you all the essentials to get you going, like visas, and tell you what opportunities are waiting to be seized!

Save the World

Do you want to make a difference while you make a dollar? From those at the frontline of global aid to the admin staff behind the scenes, you'll hear straight from real people what working for a cause is really like. Whatever your interests, whatever your skills, this book will show you how to get a job saving the world.

Coming soon!

Be Your Own Boss

If you were the kid making a profit from trading collector cards in the playground, this book is for you! Read interviews with a range of young people who all have one thing in common – they saw an idea, dared to take risks and are now running their own business. Get practical advice from concept to boardroom, encompassing funding, marketing, people management and more.

Weird and Wonderful

Want to stuff animals for a living? How about telling jokes for a job? Career FAQs *Weird and Wonderful* tells you about creative, unusual, humorous and some quite dangerous jobs that will make great conversation at parties. It's essential reading for those who don't just think outside the square, but who want to work there too!

The Sporting Arena

Whether it's your lifelong dream to win gold at the Olympics, or you just watch a LOT of sport on TV, there is a job in the sporting world that's perfect for you. Career FAQs *The Sporting Arena* includes interviews with sportspeople of all sorts about what it's really like to make a career out of pursuing your personal best. We also talk to trainers, managers, sports marketers and journalists and others, so you can learn about the possibilities and how to get started.

LEARN MORE.
EARN MORE.

If you want to get ahead in your career you need to stay ahead. **SEEK Learning** can help you upgrade your skills and qualifications through a wide range of university and training courses from Australia's best learning providers. Study what you want, where and when you want. To talk to someone about how you can get ahead, phone **1800 891 011**.

seeklearning.com.au